Lecture Notes in Artificial Intelligence 8329

Subseries of Lecture Notes in Computer Science

LNAI Series Editors

Randy Goebel
  University of Alberta, Edmonton, Canada
Yuzuru Tanaka
  Hokkaido University, Sapporo, Japan
Wolfgang Wahlster
  DFKI and Saarland University, Saarbrücken, Germany

LNAI Founding Series Editor

Joerg Siekmann
  DFKI and Saarland University, Saarbrücken, Germany

Martin Atzmueller   Alvin Chin
Denis Helic   Andreas Hotho (Eds.)

Ubiquitous
Social Media Analysis

Third International Workshops
MUSE 2012, Bristol, UK, September 24, 2012
and MSM 2012, Milwaukee, WI, USA, June 25, 2012
Revised Selected Papers

 Springer

Volume Editors

Martin Atzmueller
University of Kassel, Knowledge and Data Engineering Group
Wilhelmshöher Allee 73, 34121, Kassel, Germany
E-mail: atzmueller@cs.uni-kassel.de

Alvin Chin
Nokia, Building 2, N. 5 Donghuan Zhonglu
Economic and Technological Development Area
Beijing 100176, China
E-mail: alvin.chin@nokia.com

Denis Helic
Graz University of Technology, Institute for Knowledge Technologies
Infeldgasse 13, 8010 Graz, Austria
E-mail: dhelic@tugraz.at

Andreas Hotho
University of Würzburg, Data Mining and Information Retrieval Group
Am Hubland, 97074 Würzburg, Germany
E-mail: hotho@informatik.uni-wuerzburg.de

ISSN 0302-9743   e-ISSN 1611-3349
ISBN 978-3-642-45391-5   e-ISBN 978-3-642-45392-2
DOI 10.1007/978-3-642-45392-2
Springer Heidelberg New York Dordrecht London

Library of Congress Control Number: 2013955477

CR Subject Classification (1998): I.2.6, H.3, G.2, H.2.8, H.5, K.4.4

LNCS Sublibrary: SL 7 – Artificial Intelligence

Typesetting: Camera-ready by author, data conversion by Scientific Publishing Services, Chennai, India

Printed on acid-free paper

Springer is part of Springer Science+Business Media (www.springer.com)

Preface

Ubiquitous and social computing are creating new environments that foster the social interaction of users in several dimensions. On the ubiquitous side, there are different small distributed devices and sensors, while on the social media and social Web side there are a variety of social networking environments being implemented in a growing number of social media applications. With these, ubiquitous and social environments are transcending many domains and contexts, including events and activities in business and personal life.

Understanding and modeling ubiquitous (and) social systems requires novel approaches, methods, and techniques for their analysis. This book sets out to explore this area by presenting a number of current approaches and studies addressing selected aspects of this problem space. The individual contributions of this book focus on problems related to the modeling and mining of ubiquitous data and social media. Methods for mining, modeling, and engineering can help to advance our understanding of the dynamics and structures inherent to the respective systems integrating and applying ubiquitous social media. Specifically, we focus on detecting communities and influential users and interaction groups in social media, as well as on modeling dynamics and geographic information in ubiquitous settings. We especially focus on collective intelligence in ubiquitous and social environments. This involves frameworks for collecting appropriate data and collecting enough data, mining the data from different social media and contexts, creating models for inferring collective intelligence, and evaluating the framework to determine the accuracy of the gathered collective intelligence. In addition, different purposes and applications can then be created from the collective intelligence of social media, including issues of integrating and validating collective intelligence of social media.

The papers presented in this book are revised and significantly extended versions of papers submitted to two related workshops: the Third International Workshop on Mining Ubiquitous and Social Environments (MUSE 2012), which was held on September 24, 2012, in conjunction with the European Conference on Machine Learning and Principles and Practice of Knowledge Discovery in Databases (ECML-PKDD 2012) in Bristol, UK, and the Third International Workshop on Modeling Social Media (MSM'2012) that was held on June 25, 2012, in conjunction with ACM Hypertext in Milwaukee, USA. With respect to these two complementing workshop themes, the papers contained in this volume form a starting point for bridging the gap between the social and ubiquitous worlds. Both social media applications and ubiquitous systems benefit from modeling aspects, either at the system level, or for providing a sound data basis for further analysis and mining options. On the other hand, data analysis and data mining can provide novel insights into the user's behavior within social media systems, and thus similarly enhance and support modeling prospects.

Concerning the range of topics, we broadly consider three main themes: communities and group structure in ubiquitous social media, ubiquitous modeling, and aspects of social interactions and influence.

For the first main theme, we focus on dynamics and structural aspects of communities. We consider advanced structuring methods with respect to both social media in "How to Carve up the World: Learning and Collaboration for Structure recommendation" by Mathias Verbeke, Ilija Subasic, and Bettina Berendt. Furthermore, an approach for topological community detection is presented in "A Topological Approach for Detecting Twitter Communities with Common Interests" by Kwan Hui Lim and Amitava Datta.

For the ubiquitous modeling approaches, we included two works on methods in a ubiquitous application setting: "Using Geographic Cost Functions to Discover Vessel Itineraries from AIS Messages" by Annalisa Appice, Donato Malerba, and Antonietta Lanza describes an approach for mining itineraries from ubiquitous data. As a combination of ubiquitous and social data, "Social Media as a Source of Sensing to Study City Dynamics and Urban Social Behavior: Approaches, Models, and Opportunities" by Thiago Silva, Pedro Olmo Vaz de Melo, Jussara Almeida, and Antonio Loureiro mines social media for identifying dynamics and behavior in cities.

Concerning social interactions and influence scenarios, Derek O'Callaghan, Derek Greene, Maura Conway, Joe Carthy, and Padraig Cunningham study interactions in politically right communities in "An Analysis of Interactions Within and Between Extreme Right Communities in Social Media." The paper "Who Will Interact with Whom? A Case-Study in Second Life Using Online and Location-Based Social Network Features to Predict Interactions Between Users" by Michael Steurer and Christoph Trattner investigates online location-based social networks in Second Life. An approach for identifying influential users is presented in "Identifying Influential Users by Their Postings in Social Networks" by Beiming Sun and Vincent Ty Ng. Finally, Tarique Anwar and Muhammad Abulaish tackle the task of modeling social media, i.e., a set of Web forums into a special social graph in "Modeling a Web Forum Ecosystem into an Enriched Social Graph."

It is the hope of the editors that this book (1) catches the attention of an audience interested in recent problems and advancements in the fields of social media, online social networks and ubiquitous data and (2) helps to spark a conversation on new problems related to the engineering, modeling, mining, and analysis of ubiquitous social media and systems integrating these.

We want to thank the workshop and proceedings reviewers for their careful help in selecting and the authors for improving the submissions. We also thank all the authors for their contributions and the presenters for the interesting talks and the lively discussions at both workshops. Only this has allowed us to produce such a book.

September 2013

Martin Atzmueller
Alvin Chin
Denis Helic
Andreas Hotho

Organization

Program Committee

Martin Atzmueller	University of Kassel, Germany
Javier Luis Canovas Izquierdo	Inria & Ecole des Mines de Nantes, France
Michelangelo Ceci	Università degli Studi di Bari, Italy
Alvin Chin	Nokia, Beijing, China
Yanqing Cui	Nokia Research
Padraig Cunningham	University College Dublin, Ireland
Laura Dietz	University of Massachusetts, USA
Denis Helic	Knowledge Management Institute, Graz University of Technology, Austria
Andreas Hotho	University of Würzburg, Germany
Florian Lemmerich	University of Würzburg, Germany
Else Nygren	Uppsala Universitet, Sweden
Claudio Sartori	DEIS - University of Bologna, Italy
Giovanni Semeraro	University of Bari, Italy
Philipp Singer	Knowledge Management Institute, Graz University of Technology, Austria
Markus Strohmaier	Knowledge Management Institute, Graz University of Technology, Austria
Christoph Trattner	Knowledge Management Institute, Graz University of Technology, Austria
Maarten Van Someren	University of Amsterdam, The Netherlands
Haoyi Xiong	Broqs Engineering Consultancy
Zhiyong Yu	Institut TELECOM SudParis, France

Additional Reviewers

Aiello, Luca Maria	Pasolini, Roberto
Hou, Haixiang	Pio, Gianvito
Hu, Tianran	Pirini, Tommaso
Kibanov, Mark	Walk, Simon
Lopez-Garcia, Pablo	Wang, Zhu

Table of Contents

How to Carve up the World:
Learning and Collaboration
for Structure Recommendation

Mathias Verbeke, Ilija Subašić, and Bettina Berendt

Department of Computer Science, KU Leuven, Belgium
{mathias.verbeke,bettina.berendt}@cs.kuleuven.be,
subasic.ilija@gmail.com

Abstract. Structuring is one of the fundamental activities needed to understand data. Human structuring activity lies behind many of the datasets found on the internet that contain grouped instances, such as file or email folders, tags and bookmarks, ontologies and linked data. Understanding the dynamics of large-scale structuring activities is a key prerequisite for theories of individual behaviour in collaborative settings as well as for applications such as recommender systems. One central question is to what extent the "structurer" – be it human or machine – is driven by his/its own prior structures, and to what extent by the structures created by others such as one's communities.

In this paper, we propose a method for identifying these dynamics. The method relies on dynamic conceptual clustering, and it simulates an intellectual structuring process operating over an extended period of time. The development of a grouping of dynamically changing items follows a dynamically changing and collectively determined "guiding grouping". The analysis of a real-life dataset of a platform for literature management suggests that even in such a typical "Web 2.0" environment, users are guided somewhat more by their own previous behaviour than by their peers. Furthermore, we also illustrate how the presented method can be used to recommend structure to the user.

Keywords: Collaborative classification, social web mining, user modelling, structure recommendation, divergence measure.

1 Introduction

Nowadays, people are faced with a huge amount of data, originating from the increasing amount of emails, stored documents, or scientific publications. Just as people order their CD collection according to genre, sort books on shelves by author, or clothes by colour, it lies in the human nature to structure digital items. This is supported by the directory structure on desktop computers or tags for bookmarks, photos, and online publications. The structuring of data is a diverse process, and the grouping of a set of items into subsets can be seen as investing it with semantics: a structuring into sets of items, each instantiating a concept.

M. Atzmueller et al. (Eds.): MUSE/MSM 2012, LNAI 8329, pp. 1–22, 2013.

In addition, these concepts depend on each other: One concept is what the other is not; there is meaning not only in assigning an item to a concept, but also in *not* assigning it to another concept; and the abstract concepts co-evolve with the assignment of concrete instances to them. This feature makes an analysis of grouping more complex than an analysis just focusing on "bipartite" relations between items and concepts, on the growth of concept vocabulary over time, etc.

Since the end result of grouping will depend on the context, tasks, and previous knowledge of the "structurer", there is no overall "optimal" grouping. Grouping according to needs and knowledge is a prerequisite for the grouping to make sense to the user. Despite the personal nature of this process, people tend to be influenced by others – be it their real-life social circle, their online friends, or social recommender systems, which could steer the user (and the grouping) in a certain direction.

A good understanding of the dynamics of these structuring activities is a key prerequisite for theories of individual behaviour in collaborative settings, and it is necessary to improve the design of current and next generation (social) recommender systems [1,15]. Furthermore, it can leverage the design of mechanisms that rely on implicit user interactions such as social search [10,8]. The goal of this paper is to develop a dynamic conceptual clustering [20] that simulates this intellectual structuring process which is able to identify these structuring dynamics. The evolution of the grouping of an individual user is influenced by dynamically changing and collectively determined "guiding grouping(s)", which we will refer to as *guides*. In this paper, we investigated two types of guides. The first one is motivated by a "narrow" view of the structurer's prior grouping behaviour and experience, while the second one starts from a "wider" perspective that also takes peer experience into account in grouping decisions.

The process in which we want to transfer the structuring of one set of items to another set of items is a combination of two data mining tasks. First, we learn a model for this grouping (classification), which can be applied to structure alternative sets of items. We refer to this model as the grouping's *intension*, as opposed to its *extension*, which is the original, un-annotated grouping of the items. The first task starts with an extension of a structuring, and learns its intension as a classification model. The second task is to use the intensions of the peer groupings and apply their classifiers for prediction, to structure a new item. It starts from defining the k nearest peer groupings for a user. To decide on the k nearest peer groupings in a situation where peers group different items, we defined a novel measure of divergence between groupings that may have a different number and identities of items. Once we obtain the k nearest peers, the task is to decide on item-to-group assignment using their groupings. Based on the presence of an item in peer groupings, this decision is based either on the extension of the peer grouping (when a peer already grouped the item) or on its intension (when a peer has not grouped the item yet). By comparing the possible end groupings with the actual end grouping, we see which guide is more likely to have determined the actual process.

Our main contributions are: (a) a new data-mining approach that learns an intensional model of user groupings and uses these to group new items. The method can be used to identify structuring dynamics, which simulates an intellectual structuring process operating over an extended period of time, (b) a new divergence measure to define divergence between groupings of non-identical item sets, (c) a study of grouping behaviour in a social bookmarking system, and (d) two systems illustrating the proposed method to recommend structure to the user.

This paper is structured as follows. The following section reviews related work. Section 3 describes the method for identifying the structuring dynamics via grouping guidance, where we first outline the general system workflow and subsequently present a formalisation of the different parts. Section 4 contains an empirical analysis of this process that centres on item grouping in a real-life data set. In Section 5 we show that the grouping of new objects can also be used to recommend structure to a user, which is illustrated by means of two systems embodying this idea. Finally, Section 6 concludes the paper and presents an outlook for future work.

2   Related Work

Our work is closely related to the research on personalised information organisation, where the goal is to understand how the user can be supported in searching, exploring, and organising digital collections. This field studies how users structure data collections. Current approaches usually concentrate on individual aspects such as search (e.g. personalised ranking), rudimentary exploration support and visualisation. In [18] the authors present a taxonomy of the features that can be used for this purpose. They introduce content-based features (e.g., word distributions), content-descriptive features (e.g., keywords or tags describing textual or image content), and context-independent metadata (e.g., creator) features. The key questions from this domain that are relevant to the topic of this paper are: How is it possible to support a user in grouping collections of items?, How can we use information about the way a user groups collections to support him in structuring as-yet-unknown collections? Nürnberger and Stober [23] give an answer to this question. They describe an adaptive clustering approach that adapts to the user's way of structuring, based on a growing self-organising map. As illustrated in [3], hierarchical clustering can help to structure unseen collections based on the way a user structures her own collections. It models the structuring criteria of a user by constraints in order to learn feature weights. The goal of our approach is to examine these structuring dynamics and investigate how a user is guided by himself or his peers. This can be very useful in detecting a user's guiding preferences and obtaining an overview of the user characteristics in the overall system.

In characterisation studies, system usage patterns are studied to propose models which explain and predict user behaviour in these systems. A number of studies have investigated tagging systems. Closest to our work is the research by Santos-Neto et al. [26] on individual and social behaviour in tagging systems. They define interest-sharing between users as a measure for inferring an implicit social structure for the tagging community. In addition, they investigate whether a high level of tag reuse results in users that tag overlapping sets of items and/or use overlapping sets of tags. Sen et al. [28] study how a tagging community's vocabulary of tags forms the basis for social navigation and shared expression. They present a user-centric model of vocabulary evolution in tagging communities based on community influence and personal tendency, with an evaluation on a real system, namely the *MovieLens* recommender system. Musto et al. [22] investigate the relation between tagging based on one's own past tags or on the community's tags, abstracting from the dynamics of tag assignments.

However, the studies of tagging are limited to a "bipartite" view of the binary relations between tags and items (or the ternary ones between tags, items and users). They do not take into account the relations between tag assignments and tag non-assignments, i.e. the way in which different concepts interact structurally, for example by being in competition with each other.

Our method complements these results with insights into structuring dynamics and behavioural characteristics in collaborative settings. We focus on the way users group items and how they are influenced during this process. This can help to improve the design of future collaborative systems and leverage the design of mechanisms that rely on implicit user interactions.

From conceptual and predictive clustering [6,34], we used the key idea to a) form clusters of elements, then b) learn classifiers that reconstruct these clusters and c) apply the classifier for further result sets. Research on ontology re-use, in particular adaptive ontology re-use [30], investigates the modelling and mapping steps needed for re-using given ontologies, for whatever purpose. In contrast, we concentrate on re-use for grouping/classifying new objects, and on how to find the most suitable ontologies for re-use.

3 Grouping Guidance

This section introduces notation (Section 3.1) and describes the system's general workflow and the individual steps.

The system observes, for each user, an initial grouping and learns an initial classifier from it (Section 3.2). It then identifies, for a given user, which classifier to use next – the user's own or one determined by his peers (Section 3.3). It applies the chosen classifier (Section 3.4) and then updates it to reflect the new item grouping, which now also contains the just-added item (Section 3.5).

Steps 2, 3 and 4 are iterated on any item that the users want to structure. To analyse the structuring behaviour of each user, we compare the resulting groupings of this process to the real user groupings, which are the result of the user' structuring without computational assistance.

In Section 3.3, we present our measure of divergence between user groupings of non-identical item sets.

3.1   Notation

In this section we will introduce notational conventions for the basic concepts, for the groupings of each user's items and the classifiers learned from them, and for the time points through which the structuring evolves. Specific instantiations of these concepts can be found in Figures 1 and 2.

Basic Components. We use the following notation:

- Let U denote the set of all users (used symbols: u, v, w).
- Let T denote the set of all time points $\{0, 1, ..., t_{max}\}$, where t_{max} represents the time at which the last item arrives.
- Let D denote the set of all items (used symbol: d).

D will be used in combination with subscripts that represent users. Its superscripts represent time points. Thus $D_u^t \subseteq D$ denotes the set of all $d \in D$ already considered by $u \in U$ at $t \in T$. The item assigned to the structure by user u at t is an addition to this; it is represented by $d_u^t \in (D \setminus D_u^t)$.

Groupings and Classifiers. G and C are the (machine-induced) groupings for each user's items and the classifiers learned from them, respectively. G can be any of following:

OG **- Observed Grouping.** This represents the grouping of the user at the start of the learning phase. As will be clarified in subsection 3.2, this will be the grouping used to learn the initial classifier at the start of the cycle.

GS **- Simulated Grouping, guided by self.** This represents the grouping, at a certain point in time during the cycle, that is solely generated by the classifier of the user in question. This will be specified in subsection 3.4.

Gn **- Simulated Grouping, guided by n peers.** Just as GS, this represents the grouping at a certain point in time during the cycle, but now the grouping is guided by n peers of the user under consideration. This will be specified in subsection 3.4.

As was the case for item sets, subscripts denote users and superscripts denote time points. G_u^t is the grouping that holds at time t for user u (the set of u's item sets), i.e. a function that partitions the user's current items into groups: $G_u^t : D_u^t \mapsto 2^D$. We will refer to this as the *extensional definition* of a concept, where the definition is formed by listing all objects that fall under that definition, and thus belong to the set or class for which the definition needs to be formed.

C_u^t is the classifier (intension) learned from G_u^t, modelled as the function specifying an item's group according to this classifier: $C_u^t(d) = x$ with $C_u^t : D \mapsto 2^D$. The intensional definition is a concept that originated from logic and mathematics. In this context, an *intensional definition* is defined as a set of conditions for an object to belong to a certain class. This is done by specifying all the common properties of the objects that are part of this set, with the goal of capturing its meaning. Analogously to OG, GS and Gn for groupings, OC, CS and Cn are defined as the initial classifier and the classifiers guided by "self" and by n peers, respectively.

Time. Structuring is modelled as evolving through a series of discrete time points, where each new structuring-related activity (such as tagging a document in a social-bookmarking system) happens at some time point and leads to either the assimilation of the item into the given conceptual structure of G and C, or to a modification of the conceptual structure to accommodate the item. The conceptual structure remains stable until the subsequent activity.

We will refer to $next(t, u)$ with $next : T \times U \mapsto T$ as the first time point after t at which u structures an item, which changes the item set to $D_u^{next(t,u)} = D_u^t \cup \{d_u^t\}$. For each time point t' between subsequent considerations of items, grouping and classification remain unchanged. Formally, $\forall t' = t + 1, ..., next(t, u) :$ $D_u^{t'} = D_u^{next(t,u)} \wedge C_u^{t'} = C_u^{next(t,u)} \wedge G_u^{t'} = G_u^{next(t,u)}$.

3.2 Initial Classifier Learning

The task is to classify an item according to a set of concepts, based on its contents. We regard a classifier as an explanation why an object is assigned to a certain concept, i.e. its intensional definition. As indicated above, the opposite approach is the extensional definition, obtained by listing all objects that fall under that definition, and thus belong to the set or class for which the definition needs to be formed.

The goal is to determine intensional definitions for the user-generated groupings. Each cluster or group is then regarded as a class for which a definition needs to be calculated. Since different algorithms can be used for clustering, there are different ways in which these definitions can be calculated. These intensional definitions can then be used to assign new items to these clusters or groups. This can be seen as a classification task, since new, unclassified items need to be categorised.

3.3 Choosing the Classifier(s)

The selection of peer guides whose classifiers are used in grouping requires a measure of user similarity or divergence. Each user organises a set of items which he is interested in. Multiple peer users organise different, possibly overlapping, subsets of D. We start from the idea that similarity of items in peers' groupings indicates their interest similarity. The question that arises is: how to define similarity/divergence between groupings of non-identical item sets? Using a measure based on the overlap of items, such as Jaccard index, or mutual information would fit as the similarity/divergence measure if item sets overlap to a large extent. However, it is usually the case that peers have access (or interest) to a limited number of items, and only a small number of popular (easily accessed) items are available to all peers. To overcome this, we define a new measure of divergence between peers.

We assume that a user groups items based on their features, e.g. text for documents. For users u and v and their respective groupings G_u^t and G_v^t, we define the inter-guide measure of diversity $udiv$ as:

$$udiv(u,v) = \frac{1}{2}\Big(1/|G_u^t| \sum_{x \in G_u^t} min_{y \in G_v^t} gdiv(x,y)$$
$$+1/|G_v^t| \sum_{y \in G_v^t} min_{x \in G_u^t} gdiv(y,x) \Big) \tag{1}$$

where $gdiv(x,y)$, or inter-group diversity, is any divergence measure defined on two groups.[1]

The measure defined in Eq. 1 captures the differences between groupings by rewarding groups having shared items and groups containing similar items. The double average creates a symmetric measure.

We calculate the inter-guide divergence for all pairs of users $u,v \in U$ and create a $|U| \times |U|$ matrix M_{sim}. The M_{sim} matrix is used to select the most similar users to the user who is structuring items. For each user, we extract the corresponding row from M_{sim} and sort it to find the most similar peers, defined as the least divergent ones. From this list, we select the top n users as guide(s).

3.4 Classification

In the classification step, the identified determining classifiers of the peer users from the previous step are now used to classify the item under consideration. We distinguish two cases: in the first one, the user guides himself, so the intensional model of his own structuring is used as classifier for the new item. In the other case, the user is guided by his peers.

Self-guided Classification. If the user has not seen the item yet, we can use the intensional description of the user's current clustering to classify the new[2] instance.

At time t, the time where the new item arrives, the user's own classifier is applied: $x = CS_u^t(d)$. This gives the proposed group x based on the intensional model of the user under consideration. The item d is added to this x, which results in the new *grouping* $GS_u^{next(t,u)}$.

---

[1] If desired, a similarity measure can be calculated from this easily by appropriate normalisation: $usim(u,v) = (max_{w,z \in U} udiv(w,z) - udiv(u,v))/(max_{w,z \in U} udiv(w,z)$.

[2] Note that there is another possible case, namely when an item arrives that a user has already seen. However, we were able to disregard this case given the nature of our data. We used *CiteULike* data, which is a social bookmarking system for scientific papers, where we determined the groupings based on the taggings. As already shown in [26], only a very small proportion of the taggings are produced by the user who originally introduced the item to the system, and in general, users do not add new tags to describe the items they collected and annotated once.

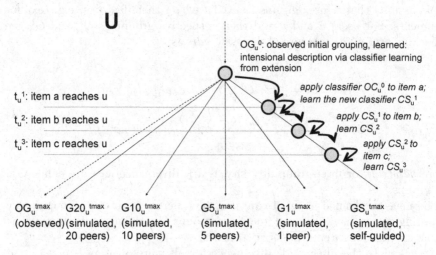

Fig. 1. General workflow for self-guided classification

This is illustrated in Figure 1, which shows the self-guided classification of a user u. It starts from the observed grouping at time point 0, OG_u^0, which is learned via classifier learning from the extensional definitions, i.e. the original grouping of user u. At each time step, when an item arrives, the previous intensional definition, i.e. classifier, from user u is used to classify the new item. After the item is added, a new intensional definition is learned. For example, when item b arrives, CS_u^1 is used to classify it, and a new classifier CS_u^2 is learned.

Peer-Guided Classification. An alternative to self-guided grouping is to use the intensional descriptions of peer users to classify the new item. We distinguish two cases: guidance by the single most similar user, and guidance by the top k peers.

Top-1 peer. If the peer user has already seen the item d, she has grouped it into one of her own groups; call this y. If she has not seen the item yet, she would classify it into her present structure into some $C1_u^t(d)$; call this y too. (Recall that the classifier is named $C1_u^t$ because the peer is u's top-1 peer at the current time.) This structuring is now projected back onto u: he will group this item into that of his own groups that is *most similar to* y, or *least dissimilar from* y, namely into his group $x = argmin_{x \in G1_u^t} gdiv(x, y)$. In both cases, d is added to the resulting x. The new grouping is represented by $G1_u^{next(t,u)}$.

This is illustrated in Figure 2, which shows the top-1 peer-guided classification of a user u, with peers v and w. It starts from the observed grouping at time point 0, OG_u^0, which is learned via classifier learning from the extensional definitions, i.e. the original groupings of user u. At each time step, when an item arrives, the most similar user is determined, and her intensional definition is used to classify the new item. After the item is added, a new intensional definition is learned.

E.g. at the arrival of item h, user v is identified as the most similar peer of user u, so consequently CS_v^1 is used to classify item h.

More than 1 peer. Let $v_1, ..., v_k$ be the top peers of u at t. For each v, a clustering of user u's grouping is calculated, with the method described for the top-1 peer. Then $y = CS_v^t(d)$ is some $y \in GS_v^t$ (v's group into which the item would be put). The result is a set of groupings, where majority voting decides on the final user group to put the item in. In case of ties, the largest $x \in Gn_u^t$ is chosen.

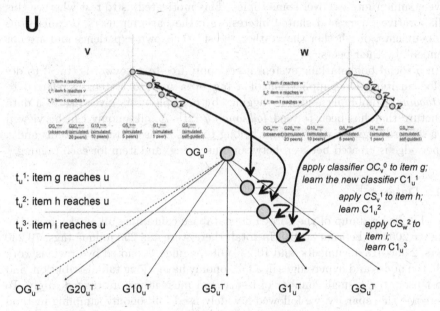

Fig. 2. General workflow for top-1 peer-guided classification

3.5 Intension Update

After the addition of the newly classified item, the intensional definition needs to be updated, which is needed to reflect the changed grouping structure implied by this addition. This is done by updating the classifier, which results in the new intensional definitions $CS_u^{next(t,u)}$, $C1_u^{next(t,u)}$ and $Cn_u^{next(t,u)}$ for the self-guided, top-1 and top-n peer-guided classification respectively.

4   Empirical Results

For testing the method outlined in the previous sections we ran a series of empirical analyses on *CiteULike*[3], a social bookmarking system designed for scientific

---

[3] http://www.citeulike.org

publications. A social bookmarking system is a perfect candidate for our grouping approach. Its Web 2.0 nature implies collaboration between peer users for enhancing the system, and by tagging, users implicitly group items. This was further motivated by the results of [26], which indicate that a rather low level of item re-tagging and tag reuse, together with the much larger number of items than tags in CiteULike, suggests that users exploit tags as an instrument to categorise items according to topics of interest. This is also indicated by results on usage patterns of collaborative tagging systems in [13]. Furthermore, they also indicate that the relatively high level of tag reuse suggests that users may have common interest over some topics. This motivated us to test whether this collaborative power and shared interests are the basis for users' organisations of documents, or whether they rather "trust" their own experience and are not "guided" by their peers.

In a social bookmarking system users apply free-text keywords (tags) to describe an item. One application of a tag by a user to a item is referred to as *tagging*. Combining multiple taggings by multiple users gives rise to a data structure that has been termed *folksonomy* [21]. A folksonomy can be viewed as a tripartite graph where nodes belong to sets of users, tags, or items and a hyper-edge is created between between a user, tag, and item for each tagging.

4.1 Dataset

We obtained a dump of the CiteULike database containing all entries from December 2004 to February 2010. In total, there were 486,250 unique tags, 70,230 users, 2,356,013 documents, and 10,236,568 taggings. As noted by previous work [13], the nodes and hyper-edges in a folksonomy have a long tail distribution, and most users tag a small number of items, and most items are tagged rarely. To overcome this sparsity, we followed a widely used folksonomy sampling method based on p-core subgraphs. The p-core subgraphs of a graph are its connected components in which the degree of each node must be at least equal to p. As in similar studies [27,16], we set the value for p to 5. Further, we constrained the dataset with regard to time and analysed only the taggings from 01/2009 until 02/2010. In total this dataset had 12,982 tags, 377 users, 11,400 documents, and 124,976 taggings. We refer to this folksonomy as F. For each document in the p-cores, we obtained the text of the abstract available on the CiteULike website.

4.2 Initial Grouping

There are structures where people group explicitly, such as mail folders, and others where people group implicitly. Tagging is an example of the latter. Since organising and structuring items is one of the largest incentives for using a social bookmarking system [28], we assume that the tagging assignments the user has at the start of the learning phase are a reasonable starting point to create the initial groupings. In accordance with Section 3.1, we use the word "observed" for these groupings, fully aware that these are generated by clustering, but we wanted to make the distinction based on the data which was used for grouping (observed data).

In order to learn those observed groupings, we split the dataset into two parts. The first part, containing the first 7 months of our dataset, is used for learning the initial groupings $G_{\bullet}^0$. On this part, we apply a clustering algorithm as in [4,3]. For a user u, we first restrict F to contain only nodes connected to u. This gives rise to a graph F^u. We then project F^u to a graph SF^u in which nodes belong to documents from set D_u^0 and edges are created between these nodes when some tag is applied to both documents. The weight of the edges in SF^u is equal to the number of tags that two documents share in F^u. We then applied a modularity clustering algorithm [33] to partition the SF^u graph. Each partition is treated as one group in G_u^0. This is repeated for all $u \in U$ to obtain initial groupings of all users, and resulted in an average of 6.59 groups per user.

The next step is to learn the initial classifiers. For this purpose, we used the Naive Bayes Classifier implementation in the WEKA data mining toolkit [14]. We used the bag of words representation of each publication's abstract as features for classification. This motivates our choice for Naive Bayes, since it has the property of being particularly suited when the dimensionality of the inputs is high. We also used a kernel estimator for modelling the attributes, rather than a single normal distribution, since this resulted in a better classification accuracy. The kernel estimator uses one Gaussian per observed value in the training data, and the conditional probability output by a kernel estimator is the weighted combination of the conditional probabilities computed from each of its constituent Gaussians.

4.3 Simulating Groupings

We represent groups belonging to a grouping using language models. At time t for a user u for every of his groups $x \in G_u^t$ we create a language model Θ_x. To find the most similar peers to a user u we calculate his inter-guide divergence (Eq. 1) to all users $v \in U$. In our experiment on social bookmarking, as the inter-group divergence $(gdiv)$ we used Jensen-Shannon divergence (JS) [7]. For two language models Θ_x and Θ_y representing groups x and y belonging to groupings $G_u^{\bullet}$ and $G_v^{\bullet}$, JS is defined as:

$$JS(\Theta_x, \Theta_y) = \frac{1}{2} KL(\Theta_x, \Theta_z) + \frac{1}{2} KL(\Theta_y, \Theta_z), \qquad (2)$$

where the probability of every word in Θ_z is the average probability in Θ_x and Θ_y; $KL(\Theta_{\bullet}, \Theta_{*})$ is Kullback-Leibler divergence between two language models.

4.4 Results

Once we obtained the groupings for GS and $G\{1|5|10|20\}$ of one user, we compared these groupings with his observed groupings at t_{max}. Like the observed initial grouping of a user u, his observed final grouping at t_{max} is a structuring of all the documents he has considered at that time. Every simulation run (whether

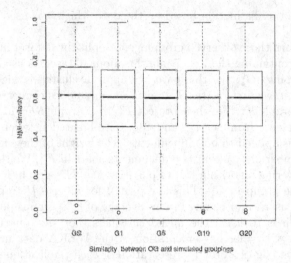

Fig. 3. Similarity between OG and simulated groupings

it be guided by self, the top-1 peer, or the top-k peers) also considers the same sequence of documents arriving for u. Therefore, all simulated groupings and the observed grouping of one user at t_{max} contain the same document set. To compare these groupings, we investigated the similarity between them. Since the groupings to compare contain the same set of documents, we can use *normalised mutual information* (NMI). It has desirable properties: it has a value between 0 and 1, and it behaves well when calculated for groupings with different numbers of groups [24].

The normalised mutual information (specifically, NMI 4 [31]) of two groupings G and G' is defined as

$$NMI(G, G') = H(G) + H(G') - \frac{H(G, G')}{\sqrt{H(G)H(G')}} \qquad (3)$$

where $H(G)$ is the entropy of grouping G and $H(G, G')$ the joint entropy of G and G' together.[4]

The similarity results for all groupings are shown in Figure 3. Our research question was to investigate whether users are guided in structuring items by their own experience or by their peers. The results suggest that in spite of the collaborative nature of a social bookmarking system, users tend to keep to their structuring system. The NMI between OG and GS is highest having a mean of 0.61 (st.dev: 0.2). Using the closest peer grouping for grouping ($G1$) does not produce more similar groupings, having a mean of 0.57 (st.dev: 0.21). Including more guides into decision does not improve the results; in our experiment the

---

[4] For groups x, y and p the distribution of items over them:
  $H(G) = -\sum_{x \in G} p(x) log_2 p(x)$ and $H(G, G') = -\sum_{x \in G} \sum_{y \in G'} p(x, y) log_2 p(x, y)$.

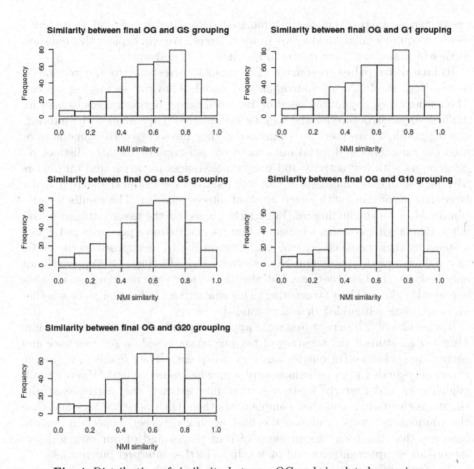

Fig. 4. Distribution of similarity between OG and simulated groupings

best peer grouping is for $G20$ (mean: 0.58, st.dev: 0.21). This is more similar to OS when compared to $G1$, but still less similar compared to GS.

Given the high standard deviation for all groupings, we investigated whether some users are more influenced by some guides. To discover this, we looked more closely into the distribution of NMI across all users. Figure 4 shows the results. We expected to see a more bimodal distribution if some users "prefer" different guides. However, all distributions fit to a normal distribution skewed to the right (towards higher similarity). Normality of the distribution is tested using the Shapiro-Wilk test, and all p values are higher than 0.05. This suggests that the majority of users behave in the same manner when it comes to structuring their items.

The results should be interpreted with some caution, and they lead to further research questions. The differences of fit were not very large and not significant. And a result that shows that on average, a user's observed grouping is 0.61 similar to his simulated self-guided result and 0.57 to the simulated peer-guided result, can mean two things: the simulated self-guided result and peer-guided result are quite similar, or they are quite different. We will explore this question further by considering our results in the light of related work.

Relation to other research. The results correspond to the results by Santos-Neto et al. [26] on individual and social behaviour in tagging system. The authors used a solution based on interest-sharing between users to infer an implicit social structure for the tagging community. They defined the interest-sharing graph, where users are connected if they have tagged the same item or used the same tags and found out that users self-organise in three distinct regions: users with low activity and unique preferences for items and tags, users with high similarity among them, but isolated from the rest of the system, and a large number of users with mixed levels of interest sharing. The results are also compatible with the findings of [13], which – based on the usage patterns in collaborative tagging systems – indicates that users are drawn primarily to tagging systems by their personal content management needs, as opposed to the desire to collaborate with others. Our results extend those findings: Our approach not only looks at the tags they use, but also at the way they use the tags vis-à-vis one another, i.e. in their structuring. This can give an indication as to whether users are more self-guided than peer-guided.

The results of [22] may at first sight appear to point into a different direction: Musto et al. studied the accuracy of tag prediction based on self-guidance and community-guidance (in our terminology: all peers). They found that (a) the community-guided tag predictions were somewhat more accurate than the self-guided ones and that (b) a strategy that first extracts tag suggestions from the items themselves and then complements them with personal tags and then the community's tags produced the best accuracy. Closer inspection reveals, however, that the basic assumptions of that model and of our own form an interesting complementarity and may help to further interpret our results.

The basic difference is that we are not interested in the tags per se, but in the mental structure that they reveal. We will explain the complementarity using a very simple fictitious example. Assume a user u and a user v (who represents the whole community). u has tagged items D_1 (some set of items) with the tag tag_1, and D_2 with tag_2. v has tagged some D_3 with tag_3 and some D_4 with tag_4. All 4 tags are different. A new item arrives for u that is most similar to his group D_1 and to the community's group D_3. If the user is guided by self, our model will predict the new item to go into D_1; if he is peer-guided, our model will predict that the new item goes *into that of u's groups that is most similar to* D_3. Now assume that D_1 is most similar to D_3. This corresponds to a situation in which everybody more or less structures in the same way. Our model will then predict that the new item goes into D_1. And the final observed grouping will be quite similar to the self-guided simulated one, and also to the peer-guided simulated one. (And this is what we found.) Musto et al., in contrast, look at the tag manifestations of this mental structure. Their model predicts for self-guidance that the new item will be tagged with tag_1, and for peer-guidance that it will be tagged with tag_3. So if vocabulary converges in the population, the item will be tagged with tag_1 *and* tag_3, and the personal-plus-community model will produce the best prediction. (And this is what Musto et al. found.) Thus, one interpretation that is consistent with the results of both studies is that while there is some more self-guidance, the differences to peers are actually quite small, and individual vocabularies do tend to converge to the community's vocabulary.

This interpretation however relies on an assumption about dynamics (the convergence of vocabulary), and dynamics were not modelled by Musto et al. Also, there are combinations of equality/difference in grouping on the one hand and vocabulary on the other, that are associated with different predictions in the two models. These combinations may persist in specific subcommunities. Finally, the levels of accuracy/NMI that they/we measured indicate that other factors play an important role too in tagging. In future work, this should be investigated in more detail.

5 Self-guided and Peer-Guided Grouping: Influence vs. Interactive Recommendations

In the previous sections, we have investigated self-guided and peer-guided grouping as a model of how content organisation in social media may evolve. This is a study of influence as it (may have) happened. However, the grouping of new objects (and the continued learning of classifiers this entails) can also be used to recommend structure to a user. In other words, we can move from a descriptive study of influence to a prescriptive suggestion of influence. We have created systems embodying this idea in two domains: the structuring of scientific literature and "friends grouping" in online social networks.

5.1 Interactive Self-guided and Peer-Guided Grouping: Reference Management

The *CiteSeerCluster* and *Damilicious* tools [5,32] help users structure a list of scientific articles. Both assume that these lists are the results of using a search engine for a query q for user u. This decision was made to allow us to integrate the tool with a real-world platform for literature search (CiteSeer), but it would be straightforward to also allow the user to upload a list obtained in some other way. To make the list more manageable, it is clustered into subgroups based on the textual contents or links to other articles, using typical clustering algorithms used and validated in bibliometrics. The user can then work with this structure, naming clusters based on proposals based on textual analysis, deleting or moving items from one cluster to another, or creating new clusters.

Each group of publications in the final grouping is regarded as the extensional definition of a concept. An intensional description of each concept is created by the top-10 TF.IDF terms in the grouped texts and, optionally, user input (CiteSeerCluster) respectively the top Lingo phrase of the group's original cluster (Damilicious). The advantage of the former is greater adaptivity, the advantage

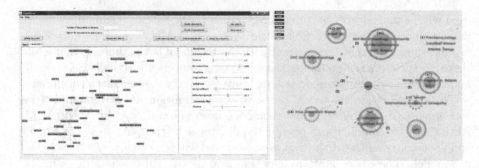

Fig. 5. Damilicious and FreeBu: groups and their elements as group-element graphs or as elements arranged "around" their groups

of the latter is better human understandability. Figure 5 (left) shows the Damilicious interface with group labels and publication IDs. (The latter can be clicked on for information and full-text access.) User u may later re-use her own classification of the results to q for a new query q', or she may use another user v's classification. This user v may be selected on different grounds, and homophily is only one of them. Thus, a user may decide to not follow the groupings of the v who is most similar to her, but to try and use the groupings least similar in order to broaden her horizon.

5.2 Interactive Self-guided and Peer-Guided Grouping: Online Social Networks

The basic idea presented in the present paper can also be applied to other domains in which people categorise and influence one another in categorising. One domain that we are exploring in current work is that of "friend grouping" in online social networks.

In environments such as Facebook or LinkedIn, people assemble large collections of contacts – other users of the same online social network who want to come into or remain in contact with the present user, often for purposes of communicating with one another, but also to maintain weak social ties. The communication may also be unidirectional, as in the *follows* relationship in Twitter, where Twitter users that one "follows" often constitute one of many information sources. Sets of contacts grow quickly, and the average Facebook user now has several hundred "friends", with counts ranging from 141.5 overall to 510 as the value for the age group 18–24.[5] Such unstructured multitudes may present problems for targeted communication and privacy management, since often messages are posted that should not really be read by everyone, and groups of contacts may be used as filters to selectively target certain recipients only. There may

---

[5] The first number results from dividing two numbers that Facebook reports in its 2013, first quarter, Financial results: the total number of friend connections divided by the total number of accounts ("users") [29]. The second number is the result of a 2000-people telephone survey [2].

also be other reasons for bringing structure into sets of contacts, for example to make specific friends' activities more visible for oneself.[6]

Current commercial online social networks support such structuring in two forms: by allowing users to create groups of contacts manually (Google+ circles, Facebook lists) or by employing a simple-looking but undisclosed classification algorithm (Facebook smart lists appear to group by the values of attributes such as "school"). The idea of using more advanced forms of data mining for deriving a good grouping structure has been pursued for a while [9,17], and the search for the best algorithm to reconstruct "ground-truth" friend-grouping data is a topic of current research [19]. However, our recent findings indicate that the functionality currently offered is barely used, and it is also questionable whether a context-free type of grouping that works along the same criteria across different people actually exists.

Thus, learning such groupings for each user individually – and maybe also for uses in different contexts – appears to be a better approach. Also, since the machine learning can never be perfect, adding interactivity, i.e. the possibility to change the assignment of people to groups, suggests itself as useful. We have implemented this in our tool FreeBu [11], see Fig. 5 (right). It uses modularity-based clustering of the user's Facebook friend graph for the initial grouping, a choice motivated by a requirements analysis and validated by a comparison with other state-of-the-art algorithms on a ground-truth data set [12].

In an interview-based evaluation of the tool, we found that users were intrigued by the tool's functionality, and were curious as to how the groupings were derived. This calls for explanation options, and one way of explaining is an intensional description of the groupings. So far, the tool uses a characterisation ("labelling") in terms of labels derived from the grouped contacts' profiles. However, participants of our user study also remarked that this often leads to unintuitive results. The conceptual-clustering idea proposed in the present paper, in which the explanation is constructed from a classifier learned from the groups, is likely to be a better solution. We have tested this with a weighted-kNN classifier and an explanation based on the group's extension plus common attributes of people in this group. We believe that such explanations functions will be key for empowering users through a better understanding of the offers of personalisation received from online social network platforms themselves or from add-on tools such as FreeBu.

So far, repeated (re-)groupings and learning have not been implemented as part of FreeBu, since we believe that a careful study of the usefulness and use of the created friends groups needs to precede such longer-term use cases, and since there are still many open questions regarding "friends management" in Facebook, the platform for which FreeBu is currently implemented. However, we expect that this will have a lot of potential, at least in online social networks that, like Twitter or LinkedIn, are treated in an "information-processing way" in which people may want group their contacts for example by information category,

---

[6] See for example Facebook's explanations of their lists,
   http://www.facebook.com/help/204604196335128/

professional context, research area, or similar, and in which they may want an explicit and user-adaptive account of how they manage these categories.

Possible social effects in grouping friends, i.e. peer-guided decisions in grouping, are another area of future work for FreeBu. The whole idea of online social networks is built on "the social", but what does this mean? People are influenced by their friends in their choices of what to disclose and how to behave in online social networks. But (how) is the very fabric of the social network, namely the accepting and rejecting of friendship requests, and/or the mental model that people have of their social surroundings, shaped by peers and their influence? Future work could create the correlational and experimental settings in which these options are prescriptive suggestions, carry out user studies to evaluate them, and also investigate how datasets could be assembled in which such influences could be studied in descriptive settings.

5.3   Evaluating Interactive Uses of Grouping Guides

Interactive uses such as the ones described in the previous two sections require a very different form of evaluation than models of influence such as the one described in the main part of this paper. These evaluations cannot be performed on historical data, but must involve users. They also need to ask more questions beyond the focus of historical-data analysis, i.e. beyond "did this influence the current user's behaviour" (or in fact "is the current user's behaviour consistent with being influenced in this way"). First, the interactive tool should be one that people like using, so a number of measures of usability and perceived usefulness are an integral part. An example of this is shown in [5]. Second, the interactive tool should ideally produce something that is indeed useful for the user's current task – to measure this goal, one must also measure other criteria like "are the groups semantically meaningful in the application domain" [5] or "do the groups support another relevant behaviour in the application domain" [12].

The user-centric evaluations we performed [5,12] suggest that the tools are a good support for their respective tasks and well-liked by users. Still, we believe that the tasks they "really" support (building up knowledge about scientific areas, communicating over a social-networking site) are more complex and longer-term than those that can be tested in the usual form of one-shot laboratory user studies, and that this should be investigated in future work [25].

6   Summary and Outlook

Summary. In this research, we investigated collaborative grouping of items and built a framework that simulates this process. Specifically, we were interested in how different users structure items depending on the influence that guides this structuring process. We developed a new method that learns and combines classifiers for item set structuring. This method starts by, in the first step, learning a model for an existing grouping, which we referred to as the intensional definition. The second step uses these learned intensions to classify new items.

To decide on the most appropriate users to take into account when grouping a new item, we defined a new divergence measure between groupings that may have different numbers and identities of elements. This method is applied to simulate the intellectual structuring process which underlies these structuring dynamics. We tested this approach on *CiteULike*, a social-bookmarking platform for literature management. The results of the study can have implications for system design of recommender systems and social search methods, for which a good understanding of the dynamics of these grouping activities is a prerequisite.

In addition to the simulation framework and its evaluation on historical data, we also described past and ongoing research on building interactive tools with which users can profit from their own and others' structuring of content and conceptual-clustering models learned from such structuring.

Limitations of the Analysis of Historical Data. The main question we addressed was one of the difference in groupings, and we did not look at the benefits users have from adopting some guides. As a baseline we used *observed groupings*, which are not explicit user groupings, but implicit groupings learned based on users tag assignments. We are aware of the possible bias here and do not claim that these are the "best" groupings users can have. Also, we used a rather simple classifier, and a limited dataset. Our current analysis provides insights both into the grouping behaviour in general and into the behaviour of users in social bookmarking systems. In future work, with a more extensive dataset, this could be extended to an iterative approach where the groupings are evaluated at different time points to evaluate the impact of including new items in the construction of the classifiers.

Challenges for Evaluation. The evaluation of grouping systems and therefore the design of better grouping systems face a number of challenges. One is the environment: If a recommender system operates in the platform itself (as is the case for example in Bibsonomy or CiteULike), then this may influence behaviour and therefore the historical data. Such influences are exacerbated in real-life applications such as reference management or social-network use, in which people also use other software such as search engines that also effectively act as recommenders. An interactive grouping system will also operate in such a pre-given environment and be affected by its recommender functionalities.

A special challenge for peer-guided grouping is privacy: an interactive grouping system divulges substantial information about the peer who is proposed as the "guide" or "donor" of a grouping. In contrast to recommender systems that recommend an item or a tag, the recommendation of a grouping will in many cases make it possible to identify the guide. This linking to a person may be a desired feature for the recipient user ("I want to group like someone who is knowledgeable"), but it may not be desired in all settings by the donor users. Thus, platforms offering peer-guided groupings should clearly describe this functionality and/or limit it to smaller groups whose participants trust each other and agree to this use of their data. These privacy concerns on the user side of course also imply privacy concerns on the data-controller side, which makes it more difficult for them to release historical datasets and for researchers to obtain such datasets for analysis.

Future work. Our method can be of use in different applications for (tag) recommendation and social search, where the grouping dynamics and behaviour adds a new level to the current individual and social measures used by these systems. Furthermore, we could extend the method to rearranging own groupings, based on the groupings of one's peers. The proposed method can also be combined with other metrics to create a hybrid measure for item set structuring.

In interactive applications, peers can be selected in different ways. One is to base the peer search on a relatedness based on personal acquaintance (as in social search), or on common properties, preferences, or behaviour (as in collaborative filtering). Alternatively, relatedness can be based on different thinking. We believe that the latter, i.e. our approach to recommending structuring, can alleviate some filter-bubble side effects on relying only on a social network and instead leverage the diversity of internet users. For example, why should the same people (= one's set of contacts in a social network) be equally well equipped to give recommendations in totally different tasks? Finding guides based on the intellectual structure of a given task and its contents allows more flexibility and therefore potentially more quality.

Conclusion. We presented a study into grouping behaviour of users. Our framework combines different data mining methods to simulate collaborative grouping of items. The results of our experiment suggest that even in such open systems as social bookmarking tools, people tend to "trust" their own experience more than turn to the wisdom of the crowd. The main question we wish to follow in the future is not one of trust, but one of the "benefit" users get by being able to choose from a myriad of diverse groupings of the items they are interested in.

References

1. Adomavicius, G., Tuzhilin, A.: Toward the next generation of recommender systems: a survey of the state-of-the-art and possible extensions. IEEE Transactions on Knowledge and Data Engineering 17(6), 734–749 (2005)
2. Arbitron and Edison Research. The infinite dial 2013. Navigating digital platforms (May 2013), http://www.edisonresearch.com/wp-content/uploads/2013/04/Edison_Research_Arbitron_Infinite_Dial_2013.pdf
3. Bade, K., Nürnberger, A.: Creating a cluster hierarchy under constraints of a partially known hierarchy. In: SDM, pp. 13–24 (2008)
4. Begelman, G., Keller, P., Smadja, F.: Automated tag clustering: Improving search and exploration in the tag space. In: Proceedings of the WWW Collaborative Web Tagging Workshop, Edinburgh, Scotland (2006)
5. Berendt, B., Krause, B., Kolbe-Nusser, S.: Intelligent scientific authoring tools: Interactive data mining for constructive uses of citation networks. Information Processing & Management 46(1), 1–10 (2010)
6. Blockeel, H., De Raedt, L., Ramon, J.: Top-down induction of clustering trees. In: Proceedings of the Fifteenth International Conference on Machine Learning, ICML 1998, pp. 55–63. Morgan Kaufmann Publishers Inc., San Francisco (1998)

7. Briët, J., Harremoës, P.: Properties of classical and quantum Jensen-Shannon divergence. Physical Review A 79, 052311 (2009)

8. Carmel, D., Zwerdling, N., Guy, I., Ofek-Koifman, S., Har'el, N., Ronen, I., Uziel, E., Yogev, S., Chernov, S.: Personalized social search based on the user's social network. In: Proceedings of the 18th ACM Conference on Information and Knowledge Management, CIKM 2009, pp. 1227–1236. ACM, New York (2009)

9. Danezis, G.: Inferring privacy policies for social networking services. In: Proceedings of the 2nd ACM Workshop on Security and Artificial Intelligence, AISec 2009, pp. 5–10. ACM, New York (2009)

10. Evans, B.M., Chi, E.H.: Towards a model of understanding social search. In: Proceedings of the 2008 ACM Conference on Computer Supported Cooperative Work, CSCW 2008, pp. 485–494. ACM, New York (2008)

11. Gao, B., Berendt, B., Clarke, D., de Wolf, R., Peetz, T., Pierson, J., Sayaf, R.: Interactive grouping of friends in osn: Towards online context management. In: 2012 IEEE 12th International Conference on Data Mining Workshops (ICDMW), pp. 555–562 (2012)

12. Gao, B., Berendt, B.: Circles, posts and privacy in egocentric social networks: An exploratory visualization approach. In: Proceedings of the 2013 IEEE/ACM International Conference on Advances in Social Networks Analysis and Mining, ASONAM 2013. ACM, New York (2013)

13. Golder, S.A., Huberman, B.A.: Usage patterns of collaborative tagging systems. Journal of Information Science 32(2), 198–208 (2006)

14. Hall, M., Frank, E., Holmes, G., Pfahringer, B., Reutemann, P., Witten, I.H.: The WEKA data mining software: an update. ACM SIGKDD Explorations Newsletter 11(1), 10–18 (2009)

15. Herlocker, J.L., Konstan, J.A., Terveen, L.G., Riedl, J.T.: Evaluating collaborative filtering recommender systems. ACM Transactions on Information Systems (TOIS) 22(1), 5–53 (2004)

16. Jäschke, R., Marinho, L., Hotho, A., Schmidt-Thieme, L., Stumme, G.: Tag recommendations in folksonomies. In: Kok, J.N., Koronacki, J., Lopez de Mantaras, R., Matwin, S., Mladenič, D., Skowron, A. (eds.) PKDD 2007. LNCS (LNAI), vol. 4702, pp. 506–514. Springer, Heidelberg (2007)

17. Jones, S., O'Neill, E.: Feasibility of structural network clustering for group-based privacy control in social networks. In: Proceedings of the Sixth Symposium on Usable Privacy and Security, SOUPS 2010, pp. 9:1–9:13. ACM, New York (2010)

18. Kashyap, V., Sheth, A.: Semantic Heterogeneity in Global Information Systems: The Role of Metadata, Context and Ontologies. In: Papazoglou, M.P., Schlageter, G. (eds.) Cooperative Information Systems, pp. 139–178. Academic Press, San Diego (1998)

19. McAuley, J.J., Leskovec, J.: Learning to discover social circles in ego networks. In: NIPS, pp. 548–556 (2012)

20. Michalski, R.S., Stepp, R.E.: Learning from observation: Conceptual clustering. In: Michalski, R.S., Carbonell, J.G., Mitchell, T.M. (eds.) Machine Learning: An Artificial Intelligence Approach, ch. 11, pp. 331–364. Tioga (1983)

21. Mika, P.: Ontologies are us: A unified model of social networks and semantics. Web Semantics: Science, Services and Agents on the World Wide Web 5(1), 5–15 (2007)

22. Musto, C., Narducci, F., Lops, P., de Gemmis, M.: Combining collaborative and content-based techniques for tag recommendation. In: Buccafurri, F., Semeraro, G. (eds.) EC-Web 2010. LNBIP, vol. 61, pp. 13–23. Springer, Heidelberg (2010)

23. Nürnberger, A., Stober, S.: User modelling for interactive user-adaptive collection structuring. In: Boujemaa, N., Detyniecki, M., Nürnberger, A. (eds.) AMR 2007. LNCS, vol. 4918, pp. 95–108. Springer, Heidelberg (2008)

24. Pfitzner, D., Leibbrandt, R., Powers, D.: Characterization and evaluation of similarity measures for pairs of clusterings. Knowledge and Information Systems 19(3), 361–394 (2009)

25. Plaisant, C.: The challenge of information visualization evaluation. In: Proceedings of the Working Conference on Advanced Visual Interfaces, AVI 2004, pp. 109–116. ACM, New York (2004)

26. Santos-Neto, E., Condon, D., Andrade, N., Iamnitchi, A., Ripeanu, M.: Individual and social behavior in tagging systems. In: Proceedings of the 20th ACM Conference on Hypertext and Hypermedia, HT 2009, pp. 183–192. ACM, New York (2009)

27. Schmitz, C., Hotho, A., Jäschke, R., Stumme, G.: Mining association rules in folksonomies. In: Batagelj, V., Bock, H.-H., Ferligoj, A., Žiberna, A. (eds.) Data Science and Classification. Studies in Classification, Data Analysis, and Knowledge Organization, pp. 261–270. Springer, Heidelberg (2006)

28. Sen, S., Lam, S.K., Rashid, A.M., Cosley, D., Frankowski, D., Osterhouse, J., Harper, F.M., Riedl, J.: Tagging, communities, vocabulary, evolution. In: Proceedings of the 2006 20th Anniversary Conference on Computer Supported Cooperative Work, CSCW 2006, pp. 181–190. ACM, New York (2006)

29. Smith, C.: By the numbers: 32 amazing Facebook stats (May 2013),
    http://expandedramblings.com/index.php/
    by-the-numbers-17-amazing-facebook-stats/

30. Stecher, R., Niederée, C., Nejdl, W., Bouquet, P.: Adaptive ontology re-use: finding and re-using sub-ontologies. IJWIS 4(2), 198–214 (2008)

31. Strehl, A., Ghosh, J.: Cluster ensembles — a knowledge reuse framework for combining multiple partitions. Journal of Machine Learning Research 3, 583–617 (2003)

32. Verbeke, M., Berendt, B., Nijssen, S.: Data mining, interactive semantic structuring, and collaboration: a diversity-aware method for sense-making in search. In: Niederee, C. (ed.) Proceedings of First International Workshop on Living Web, Collocated with the 8th International Semantic Web Conference (ISWC 2009), Washington D.C., USA, October 26, p. 8. CEUR-WS (October 2009)

33. Wan, L., Liao, J., Zhu, X.: CDPM: Finding and evaluating community structure in social networks. In: Tang, C., Ling, C.X., Zhou, X., Cercone, N.J., Li, X. (eds.) ADMA 2008. LNCS (LNAI), vol. 5139, pp. 620–627. Springer, Heidelberg (2008)

34. Ženko, B., Džeroski, S., Struyf, J.: Learning predictive clustering rules. In: Bonchi, F., Boulicaut, J.-F. (eds.) KDID 2005. LNCS, vol. 3933, pp. 234–250. Springer, Heidelberg (2006)

A Topological Approach for Detecting Twitter Communities with Common Interests

Kwan Hui Lim and Amitava Datta

School of Computer Science and Software Engineering
The University of Western Australia
Crawley, WA 6009, Australia
kwanhui@graduate.uwa.edu.au, datta@csse.uwa.edu.au

Abstract. The efficient identification of communities with common interests is a key consideration in applying targeted advertising and viral marketing to online social networking sites. Existing methods involve large scale community detection on the entire social network before determining the interests of individuals within these communities. This approach is both computationally intensive and may result in communities without a common interest. We propose an efficient topological-based approach for detecting communities that share common interests on Twitter. Our approach involves first identifying celebrities that are representative of an interest category before detecting communities based on linkages among followers of these celebrities. We also study the network characteristics and tweeting behaviour of these communities, and the effects of deepening or specialization of interest on their community structures. In particular, our evaluation on Twitter shows that these detected communities comprise members who are well-connected, cohesive and tweet about their common interest.

Keywords: Twitter, Social Network Analysis, Community Detection.

1 Introduction

Twitter is a popular micro-blogging platform that allows short messages of up to 140 characters (called tweets) to be posted and received by registered users. The popularity of Twitter is seen from its social network comprising 500 million users who produce 2,200 tweets per second [1, 2]. The popularity of Twitter and availability of data have created plenty of interest in its academic study in recent years [3–5]. In particular, this large user base and high activity level provide tremendous opportunities for companies to effectively reach out to a large group of potential customers.

One key consideration for such companies applying targeted advertising and viral marketing to online social networks is the efficient identification of communities with common interests in large social networks [6, 7]. These communities would serve as potential target audience, given their common interest (in the specific product/service). However, most of the current approaches involve first

M. Atzmueller et al. (Eds.): MUSE/MSM 2012, LNAI 8329, pp. 23–43, 2013.
© Springer-Verlag Berlin Heidelberg 2013

detecting all communities, followed by determining the interests of these communities [8, 9]. These approaches involve a lengthy and intensive process of detecting communities for the entire social network, which is growing daily. Furthermore, many of the detected communities may not share the interest we are looking for.

Our study offers a method to identify communities comprising like-minded individuals with common interests on Twitter. This method differs from existing ones that first detect all communities, followed by identifying the topics they are interested in [8, 9]. Also, our method does not unnecessarily detect communities that do not share any specific interest. Instead, our method allows for the efficient detection of only communities sharing a common interest and can be applied to targeted advertising and viral marketing (for identifying a target audience). In addition, our method is able to detect communities at different levels of interest. While there have been recent studies on detecting communities with common interest [10–12], these are interaction-based methods which use tweeting behaviour between users. On the other hand, we propose a topological-based method that uses topology links between users which are easier to collect (than the large volume of tweeting data), and also allow us to detect communities with common interest even if the users are not active in tweeting [13, 14]. Our main contributions in this chapter include the following:

- An efficient topological-based approach for detecting Twitter communities comprising users that share common interests.
- A study of the network characteristics and tweeting behaviour of Twitter communities that share common interests.
- An investigation into the effects of deepening or specialization of interest on these communities.

This chapter is structured as follows: Section 2 covers background information on Twitter; Section 3 discusses related work in the field; Section 4 describes our data and methods; Section 5 highlights our findings on community detection based on common interests; Section 6 investigates the effects of deepening or specialization of interest on these communities; and Section 7 summarizes and concludes the chapter.

2 Description of Twitter

Twitter allows registered users to post and receive short messages of up to 140 characters. These messages are called tweets and they can be posted via the Twitter website, short messaging services or third party applications. Tweets form the basis of social interactions in Twitter where a user is kept updated of the tweets of someone he/she is following. A user can also forward the tweets of others to all users following him/her, which is called retweeting. In addition, users can @mention each other in their tweets (via @username) or #hashtag keywords or topics for easy search by others (via #topic).

Twitter also provides an Application Programming Interface (API) with the functionality to collect data such as user profiles, linkages among users, tweets, retweets and @mentions [15]. This API allows developers to create applications for Twitter and researchers to study the characteristics of an online social network from the individual to community level. Currently, there is a rate limit on the number of API calls that can be executed within a specific time interval.

3   Related Work

Social networks have been intensively studied in recent years due to the availability and scale of online social networking sites. As our proposed approach aims to detect entire communities comprising users with common interests, we first discuss some related work on modeling and detecting user interests on online social networks. Next, we proceed to describe some proposed methods for detecting communities with common interests, which can be further divided into topological-based and interaction-based methods. The topological-based methods also include tag-based approaches that utilizes the tagging behaviour of users on various items to build a network graph for community detection.

3.1   User Interest Detection

One such study on user interest detection resulted in the LikeMiner system which identifies popular topics on online social networks based on the explicit "likes" indicated by users [16]. In turn, these topics can be based on textual or graphical information that are determined from comments/messages and pictures/videos respectively. LikeMiner is then able to predict the interests of a user based on the interests of his/her friends. Our approach differs from this system as we infer interest based on a user's followings instead of requiring the explicit "like" by a user. More importantly, the LikeMiner system identifies individuals whereas our approach identifies communities with common interests.

Similarly, the Friendship and Interest Propagation (FIP) model identifies interests of an individual and potential friendship links with other users [17]. The FIP model determines the interests of an individual user based on the interests of his/her friends and recommends friends based on those sharing similar interests. This model builds upon the concept of homophily which states that users with similar interests are more likely to be mutual friends compared to users with dissimilar interests. Specifically, the FIP model presents a unified framework to simultaneously identify interests and predict potential friendship links. The main difference with our method is that we identify an entire community sharing a common interest whereas the FIP model identifies an individual user's interest and recommends friendships. Also, this study was conducted on Yahoo! Pulse (pulse.yahoo.com) whereas ours is based on Twitter. Furthermore, interests are explicitly stated for the FIP model whereas our model implicitly infers interests based on a user's followings.

3.2   Topological-Based Community Detection

In their study of Twitter, Java et. al. used the Hyperlink-Induced Topic Search algorithm to detect communities based on a set of hubs and authority, and the Clique Percolation Method (CPM) to detect overlapping communities on the Twitter social network [8]. After detecting all communities, they studied the key terms used by the users (in their tweets) among these communities. Through this tweet analysis, they found that such communities share common interests, which are further divided into formal and informal ones. In addition, Java et. al. also noticed that the probability of two persons being connected is negatively correlated with their geographic distance. The difference with our approach is that we do not detect all communities then determine their interest but rather, focus directly only on communities sharing specific interests that we are interested in.

Li et. al. proposed the TTR-LDA community detection algorithm using the Latent Dirichlet Allocation model and Girvan-Newman algorithm with an inference mechanism for topic distribution [9]. They used the TTR-LDA algorithm to first detect all communities among the top 50,000 taggers in Delicious (delicious.com), followed by determining the interest topics of these communities. Next, they modeled the temporal evolution of these interest topics among the detected communities. In particular, they observed that communities share common interests which divide into defined sub-categories over time. Similar to Java et. al., they detect all communities first before determining their interest. Also, their data is based on only the top users of Delicious whereas ours is based on the full dataset of Twitter.

Using BibSonomy (www.bibsonomy.org), Atzmueller and Mitzlaff demonstrated an approach for mining communities with common descriptive features [18]. This approach integrates a database (of user attributes) and topological graph (of user links) into a dataset comprising only links connecting two users with the same attribute. Communities are then detected based on the desired attribute using this new collection of links. This approach could potentially be used to detect like-minded communities with common interests by modeling the database of user attributes as potential interests based on explicit tags on BibSonomy. While this approach can be applied to detect like-minded communities with common interest, our method is able to detect communities with varying levels of interest. We determine the interest level of users in these communities based on the number of celebrities (of a representative interest category) that these users follow. Furthermore, our method implicitly infers a user's interests based on his/her followings while the approach by Atzmueller and Mitzlaff needs to build user attributes using explicit tags on BibSonomy.

3.3   Interaction-Based Community Detection

The Highly Interactive Community Detection (HICD) method is an interaction-based approach for detecting communities where its members share a common interest and frequently interact with each other regarding this interest [10]. The HICD method uses interaction (tweeting) links to build such communities based on a threshold for their communication frequency. In the same spirit, Correa et. al. developed the iTop algorithm which uses interactions between Twitter users (@mentions and retweets) to detect topic-centric communities [11]. The iTop algorithm models the social network as a weighted graph and tries to detect the topic-centric community from this weighted graph based on the greedy maximization of local modularity. While the HICD method and iTop algorithm are able to detect interactive communities, the collection of such interaction data is a potentially time-consuming and tedious process (requiring the consistent monitoring of messages sent among such users). Our proposed approach differ mainly in the use of topological links (instead of interaction links), which is preferable when there are data collection constraints such as API call limits.

Similarly, Palsetia et. al. used interactions among Facebook and Twitter users for detecting communities comprising users with the same social interest [12]. The form of interactions used are wall posts on Facebook and tweets that mention specific Twitter users. The authors then model an undirected graph with these interactions as links and assign weights to the links based on a similarity coefficient between two users sharing a common link. Next, a modified version of the Clauset, Newman, and Moore (CNM) algorithm [19] is used in a recursive fashion to detect communities from the earlier constructed graph. Like the HICD method and iTop algorithm, this algorithm uses interaction links whereas our proposed approach uses topological links, which are smaller in volume (than interaction data) and easier to collect, especially with the stringent API call limits imposed on many online social networking sites.

4   Data and Methods

The Twitter dataset collected by Kwak et. al. [4] is used for our experimentations. This dataset was collected from 6th to 31st June 2009, comprising 41.7 million Twitter users and 1.47 billion links. In addition, the profiles of users with more than 10,000 followers are included and these profiles include details such as user ID, screen name, real name, location, etc. Kwak et. al. have made the dataset publicly available [20]. We also used the Twitter API to collect the profiles and tweets of users whom belong to either the control group or communities with common interest (as detected by our proposed approach).

We model the Twitter social network as a directed graph, $G = (U, L)$ where U refers to the set of users and L refers to the set of links. A followership link $(i, j) \in L$ indicates that user $i \in U$ is a follower of user $j \in U$, while a friendship link $Fr_{i,j}$ indicates $(i, j) = (j, i)$. We classify a Twitter user as a celebrity if he/she has more than 10,000 followers. Also, we can adjust this required number of followers to select celebrities at varying levels of popularity.

4.1   Proposed Method

Our proposed approach for detecting communities with common interest involves the following steps:

1. For a specific interest category, select a set of celebrities that represents this particular category.

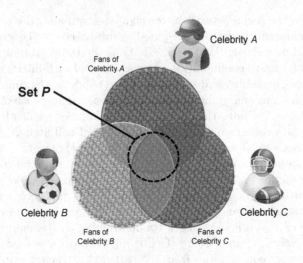

Fig. 1. Illustration of Set P

2. Based on the set of identified celebrities, select the users who follow all of these celebrities.
3. Retrieve the topology links among these users and detect communities among them.

Step 1: Representing Interest Using Celebrities. Our first step is to iden-tify a set of celebrities that represents an interest category cat, Int_{cat} and we infer the interest of a user in this category based on the number of celebrities (of cate-gory cat) that the user follows. Although Int_{cat} represents the interest level of a user in a category, this metric is subjective due to the celebrities selected. The ac-curacy of Int_{cat} is dependent on the correct classification of celebrities into their respective categories, which is subjective as some celebrities loosely belong to multiple categories (e.g. a singer that has starred in some movies). We minimize this subjective judgment by using information on Wikipedia (en.wikipedia.org) to classify these celebrities into their respective categories.

As described in [21], this process can be automated by utilizing a keyword-to-interest mapping on keywords used in either the "occupation" field or main (textual) description of the celebrity's Wikipedia article page. In particular, this keyword-to-interest mapping uses a library of 179 keywords and Table 1 gives an example of which keywords are mapped to which interest categories. This automated interest classification of a celebrity has been evaluated on a group of 1,000 celebrities with an accuracy of 83.9%. This automated process allows us to overcome the need to manually classify celebrities into their respective categories. Coupled with a secondary (manual) verification of the automated classification, we can further minimize the chances of classifying celebrities into the wrong category.

Table 1. Example of Keyword to Interest Category Mapping

Interest	Keywords
Business	entrepreneur, founder, chairman, owner, etc
Fashion	fashion designer, model, clothing designer, etc
Film & TV	actor, actress, film producer, movie director, etc
Music	singer, songwriter, dancer, band, composer, etc
Publishing	writer, author, columnist, novelist, etc

Step 2: Identifying Users with Common Interests. Our next step is to retrieve the set of Twitter users who follow all celebrities in a given category. Suppose we identify a set of n celebrities $c_1, c_2, ..., c_n$. We next identify all the followership links for the individual celebrities in this set. Consider celebrity $c_j, 1 \leqslant j \leqslant n$, and all the followership links for this celebrity $\bigcup_i link(i, c_j)$. We construct the set:

$$\mathcal{P} = \bigcap_i (\bigcup link(i, c_j)), for\ 1 \leqslant j \leqslant n$$

$\mathcal{P}$ is the set of fans who follow all the n celebrities in the set $\bigcup c_j, for\ 1 \leqslant j \leqslant n$. Fig. 1 shows an illustration of Set $\mathcal{P}$, which (in this case) are fans who follow all three sports celebrities.

Step 3: Detecting Communities Using Topology Links. For the next step of community detection, we consider only friendship links (among Set $\mathcal{P}$) for community detection as friendship links are stronger and more reflective of real-life interactions. Using this set of friendship links (which corresponds to an undirected graph), we try to detect communities among the members of $\mathcal{P}$ next using the CPM algorithm developed by Palla et. al. [22]. The CPM algorithm defines a community as one with a series of adjacent k-cliques, where a k-clique comprises k nodes that are interconnected. We first identify all k-cliques in the network and connect them if they are adjacent. Two k-cliques are adjacent if they share $(k - 1)$ common nodes. This procedure of connecting k-cliques continues iteratively until no adjacent k-cliques can be found. The result is a series of communities formed based on the k-cliques and adjacency criteria. For our experiments, we use CPM with a k-value of 3 as this produces the best results in detecting communities compared to other k-values.

Similarly, we also detect communities among the members of $\mathcal{P}$ next using the Infomap algorithm by Rosvall and Bergstrom [23]. Infomap approaches community detection as a coding or compression problem where the network graph can be compressed to retain its key structures. These key structures represent communities or clusters that are found within the network graph. Infomap uses random walks on the network graph to analyze information flow where the random walker is more likely to traverse within a cluster of nodes belonging to the

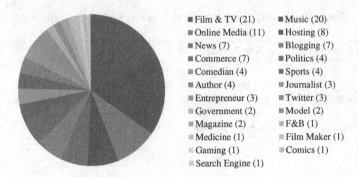

Fig. 2. Popular Categories on Twitter

same community. Using both CPM and Infomap show that our proposed method produces results that are independent of the chosen community detection algorithm and their unique characteristics, particularly in the selection of nodes that constitute the detected communities.

4.2   Experiments and Evaluation

We first study community detection and structure among individuals with a common interest in Section 5. We infer the interest of users based on the celebrities followed as users are unable to explicitly state their interests in Twitter. For this purpose, we identified six celebrities for each interest category, resulting in a total of 30 celebrities representing five categories. As a control group, we randomly chose 200,858 users to represent the group with no shared interest.[1] This control group allows us to compare the community structure of users with no common interest against users with a shared interest.

Next, we further examine how the deepening and specialization of interest affects community structure in Section 6. For this purpose, we compare communities with varying levels of interest in the specialized Country Music category against the general Music category. We selected seven winners of the Country Music Awards [24] from 2001 to 2008 as celebrities for the Country Music category based on their number of followers. Winners from 2009 onwards were not selected as the Twitter dataset [4] only comprises data until 31st June 2009. The control group in this case is the users interested in the Music category described in the previous paragraph.

---

[1] This choice of 200,858 users ensures that the control group is larger in size compared to the users with a common interest (detected using our proposed method). This control group allows us to demonstrate that our proposed method is able to detect more communities with common interests that are also larger and more cohesive compared to those in the control group, despite the control group comprising a larger number of users.

In our experiments, we measure the effectiveness of our proposed approach using the metrics of reciprocity, clustering coefficient, average path length, average degree and diameter. Reciprocity is defined as the number of friendship (bi-directional) links out of the total number of links. The clustering coefficient of a node is based on the number of triangular sub-graph that includes this node, out of all possible triangular sub-graphs. As we are interested in measuring the (entire) detected community, we take the average clustering coefficient of all nodes. In addition, we also measure the average number of links (of all nodes) and average path length (between all possible pair of nodes). Lastly, the diameter of a community (sub-graph) is based on the maximum length among all possible shortest paths. In terms of user behaviour, we also analyze the tweets posted by users in the detected communities, with a focus on their usage of #hashtags (representing their interest topics).

5   Investigating Communities with Common Interests

The Merriam-Webster dictionary defines a community as "a group of people with a common characteristic or interest living together within a larger society" [25]. Building on this definition, we propose a community detection approach based on individuals sharing common interests. We evaluate our approach by comparing the detected communities (with common interest) to our control group comprising communities with no common interest. This comparison shows that our approach of community detection based on common interests results in larger and more cohesive communities, comprising users who share common interests. Furthermore, we also show that the detected communities exhibit evidence of these common interests in the tweets they post.

For our study, we selected Film & TV, Music, Hosting, News and Blogging as categories of interest due to their popularity. These categories are selected by first identifying the top 100 celebrities based on their number of followers. Next, we used information on Wikipedia and Google[2] to determine the various categories these celebrities belong to. Following which, we build a list of categories based on the frequency of celebrities belonging to a category. Fig. 2 shows the popular categories in Twitter and we selected the five most popular categories among them.[3] For each category, we selected the six most popular celebrities based on their number of followers as listed in Table 2.[4] Also, a celebrity may belong to multiple categories (e.g. Miley Cyrus belongs to both the Music and Film & TV categories).

---

[2] If the celebrity's Wikipedia article is unavailable or not comprehensive enough, Google is used as a secondary source (e.g. news articles, fan club pages, etc).

[3] Some categories were not included due to the diversity of content within these categories (e.g. Online Commerce).

[4] Choosing six celebrities gives us an ideal number of followers (such that it is a sufficient number for us to detect meaningful communities from). While choosing a higher number of celebrities results in users with a higher level of interest, it also results in less number of followers.

Table 2. Twitter Celebrities

Screen Name	Real Name	Category
aplusk	Ashton Kutcher	Film & TV
mrskutcher	Demi Moore	Film & TV
jimmyfallon	Jimmy Fallon	Film & TV / Hosting
mileycyrus	Miley Cyrus	Film & TV / Music
PerezHilton	Mario A. Lavandeira, Jr	Blogging / Film & TV
50cent	Curtis James Jackson III	Music / Film & TV
britneyspears	Britney Spears	Music
johncmayer	John Mayer	Music
iamdiddy	Sean John Combs	Music
mileycyrus	Miley Cyrus	Film & TV / Music
coldplay	Coldplay	Music
souljaboytellem	DeAndre Cortez Way	Music
TheEllenShow	Ellen DeGeneres	Hosting
Oprah	Oprah Winfrey	Hosting
RyanSeacrest	Ryan Seacrest	Hosting
jimmyfallon	Jimmy Fallon	Film & TV / Hosting
chelsealately	Chelsea Handler	Hosting
Veronica	Veronica Belmont	Hosting
cnnbrk	CNN Breaking News	News
nytimes	The New York Times	News
TheOnion	The Onion	News
GMA	Good Morning America	News
Nightline	ABC News Nightline	News
BreakingNews	Breaking News	News
PerezHilton	Mario A. Lavandeira, Jr	Blogging / Film & TV
mashable	Mashable	Blogging
dooce	Dooce	Blogging
anamariecox	Ana Marie Cox	Blogging
BJMendelson	Brandon Mendelson	Author / Blogging
sockington	Sockington	Blogging

The next step of our community detection approach involves identifying individuals with common interests, where the interest of a user Int_{cat} is derived from the number of celebrities of category cat followed by the user. We now retrieve the list of users with $Int_{cat} > 1$, for $cat \in \{Film\&TV, Music, Hosting, News, Blogging\}$. A summary of users with $Int_{cat} > 1$ is shown in Fig. 3. In particular, we are interested in users with $Int_{cat} = 6$ as this indicates the most interest in a given category (and corresponds to users who are most interested in the product/service).

We now examine reciprocity based on link information among users with $Int_{cat} = 6$, for $cat \in \{Film\&TV, Music, Hosting, News, Blogging\}$, as shown in Table 3. Reciprocity is obtained based on the number of friendship links out of all links. The reciprocity of 15.0% to 19.6% across all categories corresponds to observations by Cha et. al. and Kwak et. al. of 10% and 22% respectively for the entire Twitter population [3, 4]. This shows that reciprocity among users with common interests is similar to reciprocity among the general population.

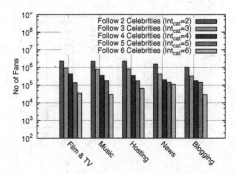 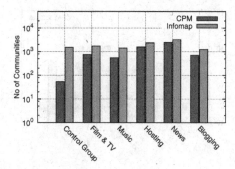

Fig. 3. Fans Following Multiple Celebrities in a Category

Fig. 4. Total Communities Detected

Table 3. Reciprocity Among Interest Groups

Category	Film & TV	Music	Hosting	News	Blogging
Reciprocity	17.9%	18.2%	15.0%	17.3%	19.6%

5.1 Analysis of Community Structure

Next, we use the CPM and Infomap algorithms to detect communities among users with $Int_{cat} = 6$, for $cat \in \{Film\&TV, Music, Hosting, News, Blogging\}$. Similarly, we detect communities among our control group comprising users with no common interest. We now compare the communities with common interests against the control group (i.e. community with no common interest) in terms of the total number of communities, size of largest community, and average community size as shown in Fig. 4, 5 and 6 respectively.

Fig. 4 and 5 show that users with common interests form more and larger communities than users without a common interest in the control group, regardless of whether CPM or Infomap was used. This is also despite the fact that the control group has a larger population of 200,858 users compared to users with a common interest, which ranges from 29,092 users ($Int_{Music} = 6$) to 109,779 users ($Int_{News} = 6$). Similarly, users with common interests form larger communities on an average as shown in Fig. 6. The exception is the News category detected using CPM as many cliques of three nodes were detected as communities thus decreasing the average community size. However, our focus is on the largest community detected as this community provides the most benefit for any application of targeted advertising and viral marketing.

The k-value chosen for CPM affects the size and number of communities detected but in all cases, we detect larger and more communities for users with a common interest compared to users without a common interest (given the same k-values). We were able to detect communities with k-values of up to 25 for the News category and we could also detect communities with k-values of 9 or higher for the other categories. For the control group, we were unable to detect

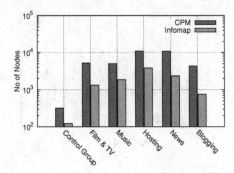
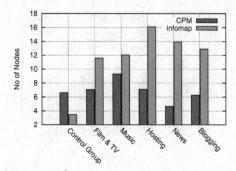

Fig. 5. Size of Largest Community Detected

Fig. 6. Average Size of Communities Detected

Table 4. Network Statistics of the Communities

Category	Control Group	Film & TV	Music	Hosting	News	Blogging
Avg. Path Length	2.83	3.03	2.82	3.09	3.35	3.09
Avg. Clustering Coefficient	0.60	0.62	0.63	0.59	0.58	0.62
Diameter	6	7	8	8	8	7
Avg. Degree	7.81	6.80	7.29	8.17	9.15	7.51

any communities at k-values higher than 6 which further proves that users with common interest form larger and more communities than users with no common interest. While the k-values affects community detection, this observation shows that our approach performs better than the control group (experiment) given the same k-values. In addition, the detection of communities at a high k-value of up to 25 also shows that our proposed approach effectively selects users who are tightly-coupled in the first place (as a k-clique is a sub-graph comprising k nodes that are fully inter-connected).

Users with common interests also form communities that are more cohesive than those without any common interest. Table 4 shows this trend where the communities with common interest have a higher clustering coefficient than our control group with no common interest, except the Hosting and News categories. However, users interested in Hosting and News have a higher average degree of links which shows that these users are better connected than users in the control group.

5.2 Analysis of Tweeting Behaviour

Apart from studying the topology structure of the detected communities (with common interests), we further evaluate their interest by analyzing their tweeting behaviour. Specifically, we study their use of #hashtags which serves as a topical label of their tweets. Using the Twitter API, we collected the most recent (last 200) tweets posted by each individual user in both the control group and detected

Fig. 7. Hashtag Cloud - Control

Fig. 8. Hashtag Cloud - Music

communities with common interests.[5] Next, we compare the usage of #hashtags in the communities with common interest against that of the control group.

Fig. 7 and 8 illustrate the #hashtag clouds for the control group and Music community respectively. From these #hashtag clouds, we observe that the control group does not tweet about a common topic, as indicated by their #hashtags which are generally unrelated and do not show a common theme (except #hotels and #travel which are related). On the other hand, the Music community tweet frequently about the Music topic, which is evident in the use of #soundcloud, #music, #nowplaying, #itunes, #grammys, etc in their tweets. While the other communities also use #hashtags about a common topic, it is to a lesser extent compared to the Music community (the main reason being that music-related topics are more popular). Another trend we observe is that all communities also display a similar interest in gaming as shown in their use of gaming-related #hashtags such as #140mafia, #mobsterworld, #gameinsight, etc.

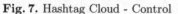

Table 5. Country Music Celebrities

Screen Name	Real Name
cunderwood83	Carrie Underwood
KeithUrban	Keith Urban
KennyAChesney	Kenny Chesney
martinamcbride	Martina McBride
paisleyofficial	Brad Paisley
TimMcGrawArtist	Tim McGraw
tobykeithmusic	Toby Keith

---

[5] While these tweets may be collected at a different time compared to Kwak et. al.'s dataset (topology links), it provides us with insight to the tweeting behaviour of these users. These tweets also show that detected communities are persistent in their common interest, despite the tweeting data being collected a few years after Kwak et. al.'s dataset.

These results show that our community detection approach detects communities that are larger, more cohesive and actively tweet about their common interests. More importantly, our approach efficiently detects communities with common interests without the need to perform large scale community detection on the entire social network. Thus, our approach is less computationally intensive (since it directly detects like-minded communities) and compares favourably to existing approaches that detect all communities then identify the interests of the communities [8, 9]. These results are also supported by observations of other authors that people with similar interests are more likely to be friends than those with dissimilar interests [26, 17].

6   Specialization and Deepening of Interests

Communities that share the same set of interests are likely to be more connected [9, 27] and interact on a more frequent basis [5]. As an extension of that argument, we show that users sharing a specialized interest form a more tightly-coupled community than users sharing a general interest. We show this by comparing users interested in the specialized category of Country Music against users interested in the general category of Music. The control group is the users interested in the general Music category as discussed in Section 5. The celebrities representing the Country Music category are seven Country Music singers who have won various awards at the Country Music Awards between 2001 to 2008 and have more than 10,000 followers. These celebrities (representing the Country Music category) are listed in Table 5.

Similar to Section 5, we used both CPM and Infomap to detect communities among users with $Int_{Country} > 1$.[6] Due to the smaller population of users following Country Music singers, the absolute number of communities detected by CPM are small (e.g. only 230 users with $Int_{Country} = 7$). We first focus on users with the most interest in Country Music, $Int_{Country} = 7$. For this user group, we detected five communities comprising 23 distinct users as shown in Fig. 9. The five communities are differentiated by nodes that are coloured green, orange, blue, yellow and purple. The grey nodes represent users that belong to multiple communities and serve as middlemen connecting the various communities. We also observed similar trends in the communities detected by Infomap.

6.1   Effects of Interest Specialization

In this section, we investigate the changes in the formation of communities and their topological structures as users specialize in their common interest (i.e. specializing in Country Music from the general Music category). To provide a relative comparison among users with $Int_{Music} = 6$ and $Int_{Country} = x$, for $2 \leq x \leq 7$, we normalize the results by the number of users in each respective group.

---

[6] We do not detect communities for users with $Int_{Country} = 1$ as this would mean all fans of any celebrity and this user group would not be meaningful for detecting communities with common interest.

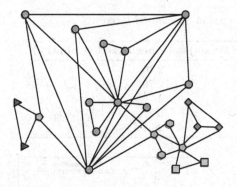

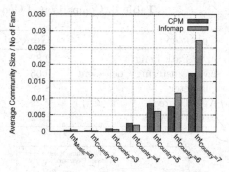

Fig. 9. Community Graph of Fans who follow all Seven Country Singers

Fig. 10. Normalized Average Community Size for Music and Country Music Categories

This normalization gives us an accurate representation of the community characteristics of each interest group without the biases of the base population size (e.g. 800 users with $Int_{Country} = 6$ compared to 29,092 users with $Int_{Music} = 6$).

The normalized average size of communities indicates the likelihood of large communities being formed among users with common interests. This measure allows us to compare if users with specialized interests form larger communities than users with a general interest. Comparing two user groups with the same level of interest in different categories (i.e. $Int_{Music} = 6$ and $Int_{Country} = 6$), we observe that the normalized average community size of the $Int_{Country} = 6$ group is 23 and 28 times larger than the $Int_{Music} = 6$ group using CPM and Infomap respectively, as shown in Fig. 10. This result shows that users sharing the same level of interest form larger communities if that interest is more specialized.

Even among users with a lower level of interest in a specialized category, they are more likely to form larger communities on average compared to users with a higher level of interest in a general category. Fig. 10 shows that users with a lower interest in the specialized Country Music category ($Int_{Country} = 3$) have a normalized average community size that is up to two times larger than that of users with more interest in the general Music category ($Int_{Music} = 6$).

Communities comprising users with a specialized interest are also more cohesive and well-connected than those with a more general interest. Table 6 best illustrates this where users with a specialized interest in Country Music form communities with a shorter average path length and diameter but higher clustering coefficient compared to those with a general interest in Music. In addition, users with $Int_{Country} = 6$ displayed a higher reciprocity of 20.1% compared to 18.2% for users with $Int_{Music} = 6$. This result shows that users with a specialized interest are more likely to be mutual followers of each other (i.e. be mutual friends) compared to users with a general interest.

Table 6. Comparison of General and Specialized Interest

Category	General (Music)	Specialized (Country)
Avg. Path Length	2.82	2.10
Avg. Clustering Coefficient	0.63	0.76
Diameter	8	4
Avg. Degree	7.29	5.52
Reciprocity	18.2%	20.1%

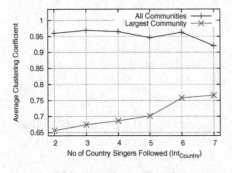 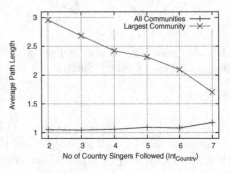

Fig. 11. Average Clustering Coefficient of Country Music Category

Fig. 12. Average Path Length of Country Music Category

6.2 Effects of Interest Deepening

Next, we investigate the changes in communities as their interest in a category grows deeper, which is indicated by an increasing Int_{cat} value. Specifically, we report on the changes in number of communities, normalized average community size, average clustering coefficient and average path length among users as their interest deepens. The size and number of communities shows how likely users with common interests form communities while clustering coefficient and path length gives an indication of connectedness within the communities.

An increase in interest level among users corresponds to an increase in their normalized average community size. Fig. 10 shows an increasing average community size with increasing $Int_{Country}$ values. This result supports our original observation that communities are more likely to be formed among like-minded individuals. In addition, the average size and number of communities formed increases as the interest level of the users increases.

Communities comprising users with common interests also get more tightly coupled as their level of interest increases. Fig. 11 shows a gradual increase in clustering coefficient among the largest communities with increasing $Int_{Country}$ values. While the average clustering coefficient of all communities remains relatively constant (from $Int_{Country} = 2$ to $Int_{Country} = 6$), this is due to the large number of small cliques detected at low $Int_{Country}$ values which increases the average clustering coefficient significantly. For example, out of 539 communities detected (with $Int_{Country} = 2$), 397 communities are cliques of three users

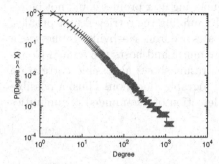

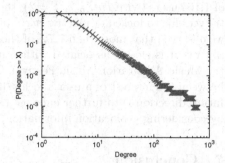

Fig. 13. Degree Complementary Cumulative Distribution Function of Largest Community with $Int_{Country} = 2$

Fig. 14. Degree Complementary Cumulative Distribution Function of Largest Community with $Int_{Country} = 4$

thus having a clustering coefficient of one. At higher $Int_{Country}$ values, less of such cliques are detected thus they have less influence on the average clustering coefficient. We are most interested in the largest community (which shows an increasing clustering coefficient) as this community has the most potential for targeted advertising and viral marketing due to its size and cohesiveness.

Fig. 12 shows an average path length of 1.7 to 3.0 hops within the largest communities at varying values of $Int_{Country}$, illustrating that users sharing common interests form communities that are better connected. This compares well with Milgram's "six degrees of separation" which states that everyone is connected by six hops of acquaintances [28]. Similarly, studies on the Microsoft Messenger social network also show that their users are separated by an average of 6.6 hops [29]. Although we compare average path length of communities and not the entire population, the largest community for $Int_{Country} = 2$ comprising 3,725 users still shows a short average path length of three hops.

These experiments show that an increasing level of interest in a category correlates with detecting larger and more communities on average. These detected communities also display characteristics of a higher clustering coefficient and shorter path length. This observation supports our initial claim that a community becomes more cohesive and tightly-coupled as its users share a deeper level of interest in a category.

The detected communities also display the characteristics of scale-free networks as shown in Fig. 13 and 14, which plots the Complementary Cumulative Distribution Function of the degree distribution of users with $Int_{Country} = 2$ and $Int_{Country} = 4$ respectively. The communities with other $Int_{Country}$ values also displayed similar trends. Upon closer examination, we observe that many individuals with large degree distribution are also country music artists but with less fans than the celebrities we have chosen (i.e. less than our threshold of 10,000 fans/followers). The fact that there are other minor country singers among these communities shows that our method effectively detects communities comprising users with a common interest. Using the Twitter API, we retrieved the profiles of 1,164 users (with $Int_{Country} = 2$), the remaining user profiles could not be retrieved due to locked or inactive accounts. Examining the retrieved user profiles, we observed that more than 7.7% of these users are from Nashville, Tennessee, a town that is closely associated with country music and hosts the annual Country Music Association Music Festival. This result shows a possible correlation between the interest of a user and his/her geographic location. Thus, a possible future direction is to further enhance the detection of like-minded communities by considering geolocation information.

7   Conclusion

In this chapter, we proposed a topological-based method to efficiently detect like-minded communities comprising individuals with common interests (Section 4), for applications in targeted advertising and viral marketing. Our method was not developed to detect all communities on Twitter. Instead, it detects larger and more cohesive communities that only comprise users who share a common interest and actively tweet about this interest. As Twitter has no explicit options for users to state their interests, we derived a measurement of interest based on the number of celebrities in an interest category that the user follows. Given the large scale and growth rate of Twitter (and other online social networking sites), our method is very scalable for identifying communities sharing common interests as it only requires topological information (and Wikipedia/Google for interest classification). The main advantage of our method is that it directly detects communities with common interests instead of having to perform a large scale community detection on the entire social network (then select communities with the common interests).

In addition, this method can also be applied to other online social networking sites by adapting to the unique characteristics of each site and their representations of celebrities and links. For example, in Facebook (www.facebook.com), celebrities could be defined as the respective Facebook pages of these celebrities and followership links as the individual user "likes" on these pages. Thereafter, our method could be applied as described in the chapter using these Facebook pages (celebrities) and user "likes" (followership links).

From a sociology perspective, we also studied the characteristics among users with a common interest compared to users without a shared interest, particularly in the way they form communities, the topological structure of these communities and their tweeting behaviour (Section 5). Also, we observed how their community structures become more connected and cohesive with deepening interest in a given category, as indicated by an increasing clustering coefficient and decreasing path length (Section 6). Similarly, the communities become more connected and cohesive as users specialize in their interest (e.g. from the general Music category to the specialized Country Music category). These observations along with our proposed method of community detection provide a tool for the implementation of targeted advertising and viral marketing, especially for products with a niche or specialized audience.

Some future areas that we are working on include the geographical analysis of communities comprising like-minded individuals and enhancing our approach to also consider geolocation data for detecting such communities. Also, we intend to perform a temporal analysis of link formation and deletion within these communities, and better understand the contributing factors of individuals joining and/or leaving communities. In addition to studying the deepening of interest based on the number of celebrities followed, we would also like to explore other definitions of deepening interest such as the celebrities' popularity (no. of followers) and the duration of this celebrity following relationship (i.e. is the user a new follower or a long-time fan).

Acknowledgments. Kwan Hui Lim was supported by the Australian Government, University of Western Australia (UWA) and School of Computer Science and Software Engineering (CSSE) under the International Postgraduate Research Scholarship, Australian Postgraduate Award, UWA CSSE Ad-hoc Top-up Scholarship and UWA Safety Net Top-Up Scholarship.

References

1. All-Twitter: Twitter to surpass 500 million registered users on wednesday. Internet (July 2012),
   http://www.mediabistro.com/alltwitter/
   500-million-registered-users_b18842
2. Engineering-Blog: The engineering behind twitter's new search experience. Internet (July 2012),
   http://engineering.twitter.com/2011/05/
   engineering-behind-twitters-new-search.html
3. Cha, M., Haddadi, H., Benevenuto, F., Gummadi, K.P.: Measuring user influence in Twitter: The million follower fallacy. In: ICWSM 2010: Proceedings of the 4th International AAAI Conference on Weblogs and Social Media, pp. 10–17 (May 2010)
4. Kwak, H., Lee, C., Park, H., Moon, S.: What is Twitter, a social network or a news media? In: WWW 2010: Proceedings of the 19th International Conference on World Wide Web, pp. 591–600 (April 2010)
5. Poblete, B., Garcia, R., Mendoza, M., Jaimes, A.: Do all birds Tweet the same? Characterizing Twitter around the world. In: CIKM 2011: Proceedings of the 20th ACM International Conference on Information and Knowledge Management, pp. 1025–1030 (October 2011)
6. Iyer, G., Soberman, D., Villas-Boas, J.M.: The targeting of advertising. Marketing Science 24(3), 461–476 (2005)
7. Kaplan, A.M., Haenlein, M.: Two hearts in three-quarter time: How to waltz the social media/viral marketing dance. Business Horizons 54, 253–263 (2011)
8. Java, A., Song, X., Finin, T., Tseng, B.: Why we Twitter: Understanding microblogging usage and communities. In: WebKDD/SNA-KDD 2007: Proceedings of the 9th WebKDD and 1st SNA-KDD Workshop on Web Mining and Social Network Analysis, pp. 56–65 (August 2007)

9. Li, D., He, B., Ding, Y., Tang, J., Sugimoto, C., Qin, Z., Yan, E., Li, J., Dong, T.: Community-based topic modeling for social tagging. In: CIKM 2010: Proceedings of the 19th ACM International Conference on Information and Knowledge Management, pp. 1565–1568 (October 2010)

10. Lim, K.H., Datta, A.: Tweets beget propinquity: Detecting highly interactive communities on twitter using tweeting links. In: WI 2012: Proceedings of the 2012 IEEE/WIC/ACM International Conference on Web Intelligence, pp. 214–221 (December 2012)

11. Correa, D., Sureka, A., Pundir, M.: iTop - Interaction based topic centric community discovery on twitter. In: PIKM 2012: Proceedings of the 5th Ph.D. Workshop on Information and Knowledge, pp. 51–58 (November 2012)

12. Palsetiay, D., Patwary, M.M.A., Zhang, K., Lee, K., Moran, C., Xie, Y., Honbo, D., Agrawal, A.: User-interest based community extraction in social networks. In: SNA-KDD 2012: Proceedings of the 6th SNA-KDD Workshop on Social Network Mining and Analysis (August 2012)

13. Lim, K.H., Datta, A.: Following the follower: Detecting communities with common interests on Twitter. In: HT 2012: Proceedings of the 23th ACM Conference on Hypertext and Social Media, pp. 317–318 (June 2012)

14. Lim, K.H., Datta, A.: Finding Twitter communities with common interests using following links of celebrities. In: MSM 2012: Proceedings of the 3rd International Workshop on Modeling Social Media, pp. 25–32 (June 2012)

15. Twitter: Twitter API. Internet (September 2011), https://dev.twitter.com

16. Jin, X., Wang, C., Luo, J., Yu, X., Han, J.: Likeminer: A system for mining the power of 'like' in social media networks. In: KDD 2011: Proceedings of the 17th ACM SIGKDD International Conference on Knowledge Discovery and Data Mining, pp. 753–756 (August 2011)

17. Yang, S.H., Long, B., Smola, A., Sadagopan, N., Zheng, Z., Zha, H.: Like like alike - Joint friendship and interest propagation in social networks. In: WWW 2011: Proceedings of the 20th International Conference on World Wide Web, pp. 537–546 (March 2011)

18. Atzmueller, M., Mitzlaff, F.: Efficient descriptive community mining. In: FLAIRS 2011: Proceedings of the 24th International Florida Artificial Intelligence Research Society Conference, pp. 459–464 (May 2011)

19. Clauset, A., Newman, M.E.J., Moore, C.: Finding community structure in very large networks. Physical Review E 70(6), 066111 (2004)

20. Kwak, H., Lee, C., Park, H., Moon, S.: Twitter dataset. Internet (June 2009), http://an.kaist.ac.kr/traces/WWW2010.html

21. Lim, K.H., Datta, A.: Interest classification of Twitter users using Wikipedia. In: WikiSym+OpenSym 2013: Proceedings of the 9th International Symposium on Wikis and Open Collaboration (August 2013)

22. Palla, G., Derényi, I., Farkas, I., Vicsek, T.: Uncovering the overlapping community structure of complex networks in nature and society. Nature 435, 814–818 (2005)

23. Rosvall, M., Bergstrom, C.T.: Maps of random walks on complex networks reveal community structure. Proceedings of the National Academy of Sciences 105(4), 1118–1123 (2008)

24. CMA: CMA Award Winners 1967-2011 (July 2013), http://www.cmaworld.com/cma-awards/winners/past-winners

25. Merriam-Webster: Merriam-webster dictionary and thesaurus. Internet (October 2011), http://www.merriam-webster.com/dictionary/community

26. Fond, T.L., Neville, J.: Randomization tests for distinguishing social influence and homophily effects. In: WWW 2010: Proceedings of the 19th International Conference on World Wide Web, pp. 601–610 (April 2010)
27. Zhao, D., Rosson, M.B.: How and why people Twitter: The role that micro-blogging plays in informal communication at work. In: GROUP 2009: Proceedings of the ACM 2009 International Conference on Supporting Group Work, pp. 243–252 (May 2009)
28. Milgram, S.: The small world problem. Psychology Today 2, 60–67 (1967)
29. Leskovec, J., Horvitz, E.: Planetary-scale views on a large instant-messaging network. In: WWW 2008: Proceedings of the 17th International Conference on World Wide Web, pp. 915–924 (April 2008)

Using Geographic Cost Functions to Discover Vessel Itineraries from AIS Messages

Annalisa Appice, Donato Malerba, and Antonietta Lanza

Dipartimento di Informatica, Università degli Studi di Bari Aldo Moro
via Orabona, 4 - 70126 Bari - Italy
{annalisa.appice,donato.malerba,antonietta.lanza}@uniba.it

Abstract. With the development of AIS (Automatic Identification System), more and more vessels are equipped with AIS technology. Vessels' reports (e.g. position in geodetic coordinates, speed, course), periodically transmitted by AIS, have become an abundant and inexpensive source of ubiquitous motion information for the maritime surveillance. In this study, we investigate the problem of processing the ubiquitous data, which are enclosed in the AIS messages of a vessel, in order to display an interpolation of the itinerary of the vessel. We define a graph-aware itinerary mining strategy, which uses spatio-temporal knowledge enclosed in each AIS message to constrain the itinerary search. Experiments investigate the impact of the proposed spatio-temporal data mining algorithm on the accuracy and efficiency of the itinerary interpolation process, also when reducing the amount of AIS messages processed per vessel.

1 Introduction

The widespread use of geo-localisation, information and telecommunication technology (e.g. Automatic Identification System - AIS) has a significant impact on maritime technology. Nowadays, a Vessel Traffic Service (VTS) is able to obtain a large volume of ubiquitous traffic information from vessels moving under its supervision. This requires intelligent maritime platforms, which should allow the VTS to manage this amount of ubiquitous data, in order to monitor maritime traffic.

The major challenge for an intelligent integrated maritime platform is the integration of a Geographic Information System (GIS) with maritime navigation systems and ubiquitous data mining services, in order to geo-localize a specific vessel, visualize both current and historic itineraries of the selected vessel and discover the vessel's behavior. In this study, we describe a data mining service for interpolating the itinerary of a vessel and tracking the vessel's motion on a selected tessellation of the world map.

The interpolation task is addressed by mining ubiquitous motion data (vessel position, speed, navigation course) which are sent periodically from the AIS. Interpolated itineraries can then be considered for further analysis, for example, to look for clusters of (near-) synchronized itineraries, which localize navy fleets, or to select the most frequent itinerary which represents the typical behavior of a vessel.

M. Atzmueller et al. (Eds.): MUSE/MSM 2012, LNAI 8329, pp. 44–62, 2013.

It is noteworthy that this interpolation task can be performed both on-line and off-line with respect to the vessel motion. In the former case, a new AIS message is interpolated when it is broadcasted. This permits the creation of a service for tracking in (near-) real time motions of a vessel. In the latter case, historic AIS messages, archived in a data warehouse, are queried to interpolate historic itineraries of a vessel. By considering that in data warehousing massive and unbounded streams of data like AIS messages are often combined with sampling techniques to reduce the amount of data stored in a server with a limited memory [3], we have the case that the interpolation service is useful when the amount of stored messages diminishes.

In general, the itinerary (or trajectory) of a mobile object is a continuous function of the time, which describes the motion of the object throughout space. In the case of a vessel, the itinerary is a piecewise function computed from the sequence of its AIS messages. The discovery of this function is an interpolation task which has received some attention in spatio-temporal data mining [11,5,12]. Nevertheless, the majority of these studies accounts for simple functions of rectilinear motions, which may not fit the non-linear motion of a vessel well (e.g. to avoid crossing land). In particular, the common practice in these studies is to process latitude, longitude and the time of consecutive AIS positions, by using a linear space-time function to compute a velocity vector and simulate the linear motion. Alternatively, velocity information is directly provided with AIS data, so that acceleration can be computed to improve the accuracy of the interpolation result.

At present, there are very few studies which investigate the problem of interpolating a non-linear motion for a vessel. They resort to complex mathematical models, such as the continuous-time Pershitz model and the discrete-time Kalman filter. They have been adapted for the interpolation of a vessel itinerary by incorporating the rate of turn, drift angle and control rudder angle deviation (see [10] for an overview). In [2] a user-decided decision rule is proposed to switch between linear and non-linear solutions.

To overcome these limitations, we formulate a spatio-temporal data mining algorithm, called ShipTracker, which resorts to a constraint-aware search strategy to address the interpolation task. In particular, it takes into account:

1. motion constraints, which are determined according to the positional, course and velocity data of the AIS messages, as well as
2. map constraints, which are imposed by a thematic (land/sea) characterization of the world map.

The algorithm operates on a world map, represented by means of the thematic tessellation model [4], which is broadly used in GIS applications. The basic unit of the space is a single (square) cell. A label is assigned to each cell, in order to describe the prominent morphology (land/sea) of the cell surface. The cell is identified by row and column coordinates in the tessellation, so that neighborhood relations between cells are implicitly defined in the tessellation matrix [1]. Since the positional data of a vessel allows us to geo-locate it into a cell, the sequence of neighbor, sea-labeled cells crossed by the vessel can be discovered as a representation of the itinerary of the vessel.

The best-first graph search algorithm is formulated to perform the interpolation of the vessel itinerary. It plugs in several cost functions for the interpolation search. In this paper, three cost functions are described and evaluated. They account for geographical-aware relations, such as the azimuth of the navigation course and the geographical distance between cells, to establish the most promising neighbor cell to be visited.

The paper is organized as follows. In Section 2, we illustrate preliminary concepts of this study. In Section 3, we illustrate the algorithm we have designed to compute and update a vessel itinerary each time a new AIS message is acquired. In Section 4, we evaluate both the accuracy and efficiency of the proposed solution, also when the amount of processed messages diminishes and finally, in Section 5, we draw conclusions.

2   Background and Preliminary Concepts

In this Section, we describe AIS data, that is, the kind of spatio-temporal data that we consider in this work. We introduce the tessellation data model that is adopted to model the world map. We illustrate some kinds of geographical-aware relations that can be used to characterize motions of a vessel. Finally, we present a way to estimate the itinerary length, based on the vessel velocity. This estimate can be used as a constraint for the itinerary search.

2.1   AIS Data

In Marine Traffic web application (http://www.marinetraffic.com/ais/) each vessel is associated to its MMSI (the unique numeric code that unambiguously identifies the vessel), name and category (pleasure craft, tug, law enforcement, cargo, tanker or other). The AIS message includes the following data:

1. the vessel MMSI;
2. the received time (day-month-year hour-minutes-seconds);
3. the latitude and longitude of the vessel position;
4. the course over ground;
5. the vessel speed.

AIS messages are sent by the transceiver installed on board the ship. It sends dynamic messages depending on the vessel speed. Messages are acquired from the Marin Traffic system and stored in a spatio-temporal database. Marine Traffic provides functionalities either to query the archive and visualize a vessel itinerary or to monitor any new movement of a vessel. In both cases, the itinerary simply moves in straight-line between the latitude-longitude point location enclosed in consecutive AIS messages, without taking into account the obstacles to be avoided.

2.2   Tessellation Model and Itinerary

In a regular tessellation of the world map, points (latitude and longitude) can be geo-located in cells of a grid having spatial resolution σ. The base unit of a map tessellation is a square (cell) $\sigma \times \sigma$ over latitude and longitude (i.e. σ the size of the cell[1]). Each cell is uniquely identified by a row index and a column index of the tessellation matrix. A label is associated to each cell to describe the morphology of the cell territory (in particular, land or sea). The motion flows through sea-labeled, neighbor-related cells. Based on these premises, the tessellation-based itinerary of a mobile object can be defined as a sequence of cells of the tessellation consdered for the world map.

Definition 1 (Itinerary). *Let us consider a tessellation of the world map with resolution σ. An itinerary with resolution σ is the sequence of neighbor cells, that is:*

$$c_1, c_2, c_3, ... c_n \tag{1}$$

where for each $i = 1 ... n - 1$, c_i is a neighbor of c_{i+1} in the tessellation matrix.

It is noteworthy that in this study the neighboring relation between cells is sought horizontally, vertically and in diagonal. When the mobile object is a vessel, only navigable (sea-labeled) cells may be included in the itinerary. The land-labeled cells host the source/destination port of a trip itinerary. Hence, they can correspond only to the starting/ending cell of the itinerary.

The data mining task for the interpolation of a vessel itinerary can be formulated by taking into account this definition of itinerary.

Definition 2 (Vessel Itinerary Interpolation). *Given*

1. *The land/sea regular tessellation of the world map.*
2. *The AIS message AIS_S sent by the vessel X at the time t_S; S is the cell of the tessellation model geo-locating the vessel position transmitted in AIS_S.*
3. *The AIS message AIS_D sent by a vessel X at the time t_D with $t_S < t_D$; D is the cell of the tessellation model geo-locating the vessel position transmitted in AIS_D.*

The goal is to interpolate the itinerary (see Definition 1) of X from S to D over the tessellation of the world map.

2.3   Geographical Distance

The geographical distance is the distance measured along the surface of the Earth. In general, the distance between points, which are defined by geographical coordinates in terms of latitude (la) and longitude (lo), is an element in solving the second (inverse) geodetic problem. Following the main stream of research

---

[1] USGS (http://edc2.usgs.gov/glcc/glcc.php) distributes a tessellation of the world map having $\sigma = 0.00833° \approx 1Km$. We use this tessellation in the experiments described in Section 4.

in several GISs, we resort to the simple spherical law of cosines to approximate geographical distance between two points.

$$d(P_S, P_D) = R \cdot \text{acos}(\sin la_S \cdot \sin la_D + \cos la_S \cdot \cos la_D \cdot \cos(lo_D - lo_S)), \quad (2)$$

where $P_S = (la_S, lo_S)$, $P_D = (la_D, lo_D)$ and $R = 6371$ km is the Earth's radius.

The geographical distance between two cells of the tessellation model is computed as the geographical distance between the geographical coordinates of the centroids in these cells.

2.4 Azimuth

The azimuth is an angular measurement in a spherical coordinate system. It is calculated by perpendicularly projecting the vector from an observer (origin) to a point of interest on a reference plane and measuring the angle between it and a reference vector on the reference plane. It is usually measured in degrees (^{o}). The concept is used in many practical applications, including navigation where it is denoted with α and defined as a horizontal angle measured clockwise from any fixed reference plane or easily established base direction line. In a general navigational context, this base direction line is typically true north, measured as a 0° azimuth. Azimuth is used to define the navigation course. We can also use the concept of the azimuth to measure the angle of the straight line connecting two geographical points with respect to the north line.

$$\alpha(P_S, P_D) = \text{atan} \frac{\cos la_D \cdot \sin(lo_D - lo_S)}{\cos la_S \cdot \sin la_D - \sin la_S \cdot \cos la_D \cdot \cos(lo_D - lo_S)}. \quad (3)$$

2.5 Itinerary Length Estimate

Let AIS_S and AIS_D be two consecutive AIS messages sent by a vessel. We formulate the hypothesis that the vessel moves with uniform acceleration between consecutive messages. This hypothesis is acceptable since, in most vessel operational conditions, the itinerary of a vessel is slowly speed changing. Under this hypothesis, physical quantities like acceleration, velocity and time can be used to retrieve a reasonable estimate of the length of the itinerary from AIS_S to AIS_D. This is done by using the classical Equation of motion as follows:

$$l(AIS_S, AIS_D) = v_S \cdot \Delta T + \frac{1}{2}a \cdot \Delta T^2, \quad (4)$$

where:

1. $v_S = speed(AIS_S)$,
2. $\Delta T = (time(AIS_D) - time(AIS_S))$,
3. $a = \frac{\Delta V}{\Delta T}$ with $\Delta V = (speed(AIS_D) - speed(AIS_S))$.

By considering that the geographical distance computed between the centroids of the cells, which geo-locate both AIS_S and AIS_D, is a lower-bound of the length of the itinerary between the same cells, the maximum itinerary length LMAX () is then estimated as follows:

$$LMAX(AIS_S, AIS_D) = \epsilon \cdot \begin{cases} l(AIS_S, AIS_D) & if \\ & l(AIS_S, AIS_D) \geq d(AIS_S, AIS_D) \text{ , (5)} \\ d(AIS_S, AIS_D) & otherwise \end{cases}$$

where ϵ is a user-defined correction factor ($\epsilon \geq 1$, by default $\epsilon = 1.5$) and $d()$ is the geographical distance computed, according to Equation 2, between the centroids of cells geo-locating AIS_S and AIS_D, respectively.

It is noteworthy that the introduction of the correction factor ϵ in Equation 5 allows us to soften the effect of underestimating LMAX when the hypothesis of a motion with uniform acceleration is no longer valid, for example, when the vessel first accelerates and then decelerates between the two messages.

3   ShipTracker

Let us consider a vessel that sends AIS messages to Marine Traffic by the transceiver installed on board the ship. Let S be the cell, which geo-locates the previous AIS message and D be the cell, which geo-locates the current AIS message. The discovery process is triggered to interpolate the unknown part of the itinerary (see Definition 1) from S to D.

ShipTracker resorts to a completely connected graph structure to represent the tessellation model of the world map. The cells are the graph nodes and the neighbor relations between cells are the graph edges. The interpolation task is then reformulated as the task of looking for a path of the tessellation-based graph, which links the nodes that geo-locate two consecutive AIS messages sent by a vessel. Several alternative paths may connect two nodes in a graph. The peculiar challenge of a path search approach is basing the interpolation search on the knowledge that we have to describe the unknown itinerary.

In ShipTracker, the search procedure is constrained by the estimate of the itinerary length (computed as reported in Section 2.5) and the land/sea characterization of the tessellation model of the map (described in Section 2.2). Cost functions are used to define the priority according to the cells which are visited. The definition of a cost function is the crucial point of the discovery process. It is plausibly influenced by several criteria, such as the distance from the destination and/or the navigation course.

3.1   Itinerary Search

The top-level description of the main routine of the interpolation process performed by ShipTracker is reported in Algorithm 1. The algorithm inputs:

1. the last visited cell C (seed of the search);

2. the destination cell D geo-referencing the new AIS message;

3. the current itinerary i explored from S to C;

4. the estimated length L of the unknown itinerary i from C to D;

5. the cost function h according to the neighbors of a cell which are sorted.

The search is performed by resorting to a best first search, which recursively visits neighbors of C until D is reached and the itinerary i is completely populated. The algorithm returns false values when the search fails and no itinerary is completed.

Initially, $C = S$, $i = \lambda$ (the empty itinerary) and $L = LMAX(AIS_S, AIS_D)$ (according to Equation 5). When C coincides with D, it is added to i and the search stops (lines 1-3, Algorithm 1). When $L < 0$, that is, the length of the itinerary explored from S to C is over-sized with respect to the estimation, the algorithm backtracks, in order to explore the alternative itineraries (lines 4-5, Algorithm 1). Otherwise, the neighborhood of C is computed (line 7, Algorithm 1). When the neighborhood is not empty, neighbors are sorted according to their priority (line 11, Algorithm 1), C is added to the itinerary i (line 12, Algorithm 1) and the sorted neighborhood is recursively explored (lines 13-18, Algorithm 1).

The neighborhood of C is the set of sea-labeled cells, which surround C in the tessellation model of the world map. For each cell, Algorithm 1 looks for horizontal and vertical neighbors, as well as for neighbors in diagonal. The priority of a neighbor is established according to the cost function h (details in Section 3.2).

In the exploration of the neighborhood, we follow the main idea in A* [7]. We expect that a vessel traverses a cell by following the itinerary of the lowest known length. This principle allows us to keep, for each cell visited in the map, a sorted priority queue of the alternate itinerary segments, which have been explored along the way to arrive at the cell itself. Then we can avoid visiting a neighbor cell when we are following an itinerary, which is longer than an alternate itinerary encountered up to the same cell in the past. In this case, the cell is traversed at the lower-length itinerary segment instead.

In the neighborhood exploration, the main procedure $itinerary()$ is used to explore itineraries, starting from the unvisited neighbors of the current seed. Let N be the unvisited neighbor of C, d be the geographic distance between C and N ($distance(C, N)$, line 14, Algorithm 1). $itinerary()$ is recursively used by specifying N as the seed for the search and $L - d$ as the estimate of the length of the unknown itinerary from N to D (line 15, Algorithm 1). If the recursive exploration of the neighborhood fails to find the itinerary when starting from the currently selected neighbor (i.e. $itinerary(N, D, i, L-d, h)$ returns false), the search process continues by visiting the remaining neighbors of C until the goal is reached (loop at lines 13-18, Algorithm 1). If the exploration phase fails for each neighbor of C, the visit procedure removes C from i ("backtracks") as soon as it determines that C cannot complete a valid solution (line 20, Algorithm 1).

Algorithm 1. function itinerary(C, D, i, L, h) **return** Boolean

Require: C {the current cell}
Require: D {the destination cell}
Require: L {estimated length of itinerary from C to D}
Require: h {cost function}
Ensure: i {itinerary}
1: **if** $C = D$ **then**
2:    i.add(C);
3:    **return** true
4: **else if** $L \leq 0$ **then**
5:    **return** false
6: **else**
7:    Neighborhood=nearby(C)
8:    **if** empty(Neighborhood) **then**
9:       **return** false
10:   **else**
11:      sort(Neighborhood, h)
12:      i.add(C)
13:      **for all** N:Neighborhood **do**
14:         d=distance(C, N) {the distance is geographically computed according to Eq.2}
15:         **if** itinerary($N, D, i, L - d, h$) **then**
16:            **return** true
17:         **end if**
18:      **end for**
19:   **end if**
20:   i.remove(C) {backtrack}
21: **end if**

3.2 Cost Function

We have considered several cost functions to assign a priority to the neighbors of a cell. The goodness of a cost function cannot be theoretically established, but it will depend on how accurately and quickly it will allow us to determine the real vessel itinerary. We expect that the goodness of a cost function would be influenced by the shape of the searched itinerary. Our idea is that several cost functions can be combined in a single one, to balance their separate drawbacks and advantages.

Azimuth-Based Cost (H1). The definition of the cost function $H1()$ is founded on the idea that, in the absence of obstacles, a vessel will tend to move following the direction of the straight-line connecting the current cell to the destination cell.

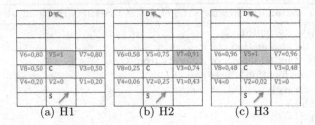

Fig. 1. Examples of ranking of neighbors of a cell, based on a different definition of the cost function

Let C be the currently visited cell, N be a visitable neighbor of C, D be the destination cell in the itinerary. Then,

$$H1(C,N) = 1 - \frac{|\alpha(C,N) - \alpha(C,D)|}{180}, \qquad (6)$$

where $\alpha()$ is computed according to Equation 3.

H1(C,N) has values in the range $[0,1]$. The higher value is assigned to the cell, which is the best candidate for inclusion in the itinerary (see Figure 1(a)).

Course-Based Cost (H2). The definition of the cost function $H2()$ takes into account the navigation course (azimuth). It is based on the idea that, if there is a substantial drift between the navigation course at the source cell and the navigation course at the destination cell, then the visited cells of the itinerary should fit this change of course in the motion.

Let C be the currently visited cell, N be a visitable neighbor of C, S be the source cell of the itinerary and D be the destination cell of the itinerary. By considering that the navigation course represents the azimuth of the velocity vector at the time of the AIS message, we can compute:

1. The absolute difference HCS computed between the navigation course of the AIS message at the source S and the azimuth of the straight line vector, connecting the current node C to its neighbor N, that is,

$$HCS(C,N) = 1 - \frac{|course(AIS_S) - \alpha(C,N)|}{180}. \qquad (7)$$

2. The absolute difference HCD computed between the navigation course of the AIS message at the destination D and the azimuth of the straight line connecting the current node C to N, that is,

$$HCD(C,N) = 1 - \frac{|course(AIS_D) - \alpha(C,N)|}{180}. \qquad (8)$$

The cost function H2() is the inverse distance weighted sum of both $HCS(C, N)$ and $HCD(C, N)$, that is,

$$H2(C, N) = \frac{HCS(C, N) \cdot \frac{1}{d(S,N)} + HCD(C, N) \cdot \frac{1}{d(N,D)}}{\frac{1}{d(S,N)} + \frac{1}{d(N,D)}}. \tag{9}$$

where $d()$ is the geographical distance computed according to Equation 2.

H2(C,N) ranges in the interval $[0,1]$. The higher value is assigned to the cell, which is the best candidate for inclusion in the itinerary (see Figure 1(b)).

Distance-Based Cost (H3). The definition of the cost function $H3()$ assigns higher priority to the neighbor cell, which is the closest to the destination.

$$H3(C, N) = \frac{|d(N, D) - d_{min}(C)|}{d_{max}(C) - d_{min} C}, \tag{10}$$

where:

$$d_{max}(C) = \max_{N \in Neighborhood(C)} (d(N, D)) \tag{11}$$

$$d_{min}(C) = \min_{N \in Neighborhood(C)} (d(N, D)). \tag{12}$$

H3(C,N) ranges in the interval $[0,1]$. The higher value is assigned to the cell, which is the best candidate for inclusion in the itinerary (see Figure 1(c)).

3.3   Which Cost Function?

By considering the pool of cost functions defined above, cells can be ranked into a neighborhood by using a function $h()$ that is formulated with the additive schema reported in the following:

$$h(C, V) = \frac{w1 \cdot H1(C, N) + w2 \cdot H2(C, N) + w3 \cdot H3(C, N)}{w1 + w2 + w3}, \tag{13}$$

where w_i weights the contribution of $H_i()$ to the final cost $h()$. In this study, we assign equal weights to the combined costs.

3.4   Example

The definition of h can influence the shape of the itinerary, as well as the number of cells visited, to output the itinerary and, consequently, the efficiency of the interpolation process. This is observable in Figure 2. In particular, the use of H2() allows us to automatically detect the non-linear form of the itinerary, while the combination of H1, H2 and H3 permits us to speed-up the entire search process by minimizing the number of cells visited to discover the final itinerary. These considerations will be confirmed by the results of the experimental study, which are illustrated in Section 4.

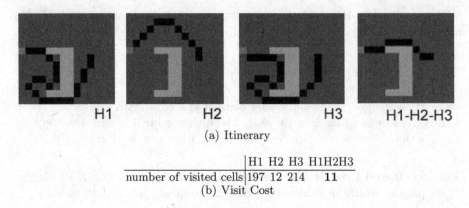

(a) Itinerary

	H1	H2	H3	H1H2H3
number of visited cells	197	12	214	**11**

(b) Visit Cost

Fig. 2. The itinerary interpolated by using h()=H1(), h()=H2(), h()=H3() and h()=(H1()+H2()+H3())/3. The navigation course at the source (red cell) is $45°N$, while the navigation course at the destination is $135°SW$.

4   Experiments

ShipTracker is written in Java. A set of experiments is run, in order to investigate the effectiveness of the itinerary interpolation process performed by ShipTracker. We run experiments on an Intel(R) Core(TM) 2 DUO CPU $E4500$ @2.20GHz, with 2.0 GiB of RAM Memory, running Ubuntu Release 11.10 (oneiric) Kernel Linux $3.0.0 - 12 - generic$.

4.1   Itinerary Description

We use the land/sea tessellation of the world map, which is distributed by USGS (http://edc2.usgs.gov/glcc/glcc.php). This tessellation is a grid of $0.00833° \times 0.00833°$ (latitude and longitude) cells ($\sigma = .00833°$, where $.00833° \approx 1km$). We also consider AIS messages for eight vessels and interpolate full itineraries for selected vessels (according to Definition 1). The speed of selected vessels ranges between 17 knots and 22 knots when vessels are in the open water.

The real itinerary of each vessel is determined based on AIS messages displayed by Marine Traffic. These messages are those broadcasted between 2-12 minutes. We consider a linear motion between AIS messages of Marine Traffic and determine the sequence of cells (real itinerary), which are crossed by the vessel. For every cell of this itinerary, we generate an AIS message. This simulation is done under the hypothesis of uniform acceleration between consecutive AIS messages of Marine Traffic. The acceleration is defined by the rate of change of velocity with respect to time. Based on these premises, real itineraries are completely specified. The length (number of cells) of the itineraries is reported in Table 1. A new cell of a real itinerary of this study is crossed between 30 seconds-two minutes, depending on the vessels' speed.

Table 1. Length (number of cells) of each itinerary by considering the tessellation of the world map provided by USGS (http://edc2.usgs.gov/glcc/glcc.php), with cell size $\sigma = .00833°$

Vessel	Itinerary length
X1	335
X2	918
X3	110
X4	237
X5	681
X6	187
X7	50
X8	42

4.2 Goals

The twin goal of this empirical investigation is:

1. To investigate how much the itinerary interpolation process is influenced by the cost function and sketch guidelines to assist the user in the formulation of the cost function h().
2. To investigate how much the itinerary interpolation process is influenced by the number of AIS messages sent by the vessel. Our goal is to prove that we can save the cost of data warehousing by diminishing the number of stored AIS messages, without compromising too much the accuracy of the itinerary, which can be interpolated by the traffic control system.

4.3 Measures

We evaluate the accuracy of the itinerary interpolated by ShipTracker and the computational cost of the interpolation process.

The accuracy is measured as the dynamic distance warping [8] (DTW) computed between the itinerary and the interpolated itinerary. We base the computation of DTW on the implementation provided in the Weka toolkit[6].

The computational cost is measured as the interpolation time (in milliseconds). Both measures are computed for each sub-itinerary, which connects two consecutive AIS messages sent by a vessel. Local measures are then averaged on the entire itinerary. In particular, let us consider:

1. a vessel X;
2. a pair of AIS messages AIS_i and AIS_{i+1} sent from X at the consecutive time points t_i and t_{i+1}, respectively;
3. the real itinerary $i_R(i, i+1)$ of X between t_i to t_{i+1};
4. the itinerary $i_S(i, i+1)$ interpolated from AIS_i and AIS_{i+1}.

We compute the average computation time $(\overline{time})$ and the average dtw $(\overline{dtw})$ as follows:

Table 2. h() function defined according to the additive schema reported in Equation 13.

experiment code	w1	w2	w3
H1	1	0	0
H2	0	1	0
H3	0	0	1
H1H2	1	1	0
H1H3	1	0	1
H2H3	0	1	1
H1H2H3	1	1	1

$$\overline{time}(X) = \frac{\sum_{i=1}^{n-1} time(i_S(AIS_i, AIS_{i+1}))}{n-1}, \tag{14}$$

$$\overline{dtw}(X) = \frac{\sum_{i=1}^{n-1} dtw(i_S(AIS_i, AIS_{i+1}), i_R(AIS_i, AIS_{i+1}))}{n-1}, \tag{15}$$

where:

1. n is the total number of AIS messages sent from X;
2. $time(i(AIS_i, AIS_{i+1}))$ is the computation time spent to complete the interpolation of $i_S(AIS_i, AIS_{i+1})$, when AIS_{i+1} is acquired;
3. $dtw(i_S(AIS_i, AIS_{i+1}), i_R(AIS_i, AIS_{i+1}))$ is the dynamic time warping computed between $i_S(AIS_i, AIS_{i+1})$ and $i_R(AIS_i, AIS_{i+1})$.

The lower the $\overline{time}(X)$, the more efficient the interpolation process and the refresh of the itinerary on the map when a new AIS massage is acquired. The lower the $\overline{dtw}(X)$, the closer the itinerary visualized on the map to the real itinerary of the vessel.

4.4   Experimental Settings

We consider several experimental settings, which are obtained by varying the cost function and the percentage of cells, randomly selected along the itinerary, to geo-locate the AIS messages mined by ShipTracker. The definition of the cost function h varies according to the schema in Table 2. The number of AIS messages ranges between 5% and 10% of the length of each real itinerary. The stream of messages is simulated. When a message arrives the sub-itinerary followed by the vessel from the last AIS message to the new one is determined. The visualization of the itinerary of the vessel is refreshed.

4.5   Results

In this Section, we illustrate the interpolations performed by varying the cost function and the amount of AIS messages processed for the interpolation.

Table 3. Cost function (interpolation accuracy): itineraries interpolated by varying the definition of h() according to Table 2 when 10% of cells of the real itinerary is chosen to geo-locate an AIS message. (Columns 2-8) $\overline{dtw}$ computed on the set of eight vessels in this study. (Column 9) Wilcoxon Matched-Pairs Signed-Ranks Test comparing the dtw computed for each local sub-itinerary interpolated with H1H2H3 and the corresponding sub-itinerary interpolated with the cost function (bestH corresponds to the results in bold), chosen vessel by vessel. (-) means that bestH is statistically better than H1H2H3 with p-value=0.1. (=) means that they perform equally well.

Vessel	h							Wilcoxon Test
	H1	H2	H3	H1H2	H1H3	H2H3	H1H2H3	H1H2H3 vs bestH
X1	**0.35**	1.19	0.62	0.63	0.58	0.64	*0.53*	(-)
X2	**0.17**	0.42	0.29	0.27	0.24	0.27	*0.21*	(=)
X3	0.65	0.92	0.69	**0.45**	0.71	0.53	*0.52*	(=)
X4	**0.19**	0.49	0.24	0.27	0.22	0.26	*0.22*	(=)
X5	**0.37**	0.71	0.47	0.48	0.44	0.45	*0.40*	(=)
X6	0.58	0.80	0.83	**0.53**	0.80	0.70	*0.54*	(=)
X7	**0.74**	0.89	**0.74**	0.84	**0.74**	**0.74**	**0.74**	(=)
X8	**0.63**	1.17	1.14	1.02	0.97	0.94	*0.73*	(=)
avg	**0.45**	0.82	0.62	0.56	0.58	0.56	*0.48*	

Cost Function Study. We vary h() as reported in Table 2. For each full itinerary, we randomly select 10% of the cells in the itinerary and use selected cells, in order to geo-locate AIS messages interpolated by ShipTracker.

The results of the accuracy of interpolated itineraries ($\overline{dtw}$) are collected in Table 3, while the results of the efficiency of interpolation process ($\overline{time}$) are reported in Table 4. Figure 4.5 displays the itinerary interpolated for the vessel X2 when the cost function is H1H2H3.

The evaluation of the accuracy of the interpolation process shows that, when comparing H1, H2 and H3 (columns 2-4 of Table 3), H1 discovers the more accurate itinerary, while H2 discovers the less accurate itinerary. H3 is in the middle. If we extend this analysis to the combined costs H1H2, H2H3, H1H3 and H1H2H3 (columns 5-8 of Table 3), we observe that H1H2 retrieves the itinerary that fits at the best the real one when the vessel frequently changes the course of its navigation (e.g. X3 and X6). This suggests that H2, used in combination with H1, can automatically fit the itinerary to the non-linear course of navigation without the need for a decision rule specified by the user. By completing this analysis, we observe that the cost H1H2H3, which combines both H1, H2 and H3, is always more accurate after H1 (or H1H2) for all vessels in this study.

The evaluation of the efficiency of the interpolation process shows that H1H3H3 is also the most competitive cost function in terms of computation time (Table 4). The interpolation time is about 1 milliseconds per message. This result is obtained by exploiting the geographic relational information enclosed in the map (the navigation proceeds by crossing neighboring sea cells which are visitable with a shorter itinerary) and in the AIS data (the navigation proceeds towards cells with the highest priority, based on the geographical distance from the destination and

Table 4. Cost function (interpolation time): itineraries interpolated by varying the definition of h() according to Table 2 when 10% of cells of the real itinerary is chosen to geo-locate an AIS message. (Columns 2-8) $\overline{time}$ (in milliseconds) is computed on the set of eight vessels in this study. Computation times are obtained by running ten trials of the itinerary interpolation algorithm and computing the average time on these trials. The lower time is in bold.

Vessel	H1	H2	H3	H1H2	H1H3	H2H3	H1H2H3
X1	1.15	36.74	**0.97**	1.77	**0.97**	1.00	0.99
X2	2.22	9.16	1.80	3.47	1.85	1.85	**1.78**
X3	1.60	57.4	1.4	0.80	0.80	0.90	**0.80**
X4	4.36	7.18	1.05	1.27	1.27	1.36	**1.04**
X5	2.66	23.53	1.39	2.97	1.66	1.34	**1.32**
X6	1	35.7	0.89	2.84	0.84	0.84	**0.83**
X7	4.8	5	2.6	3.4	1.4	2.4	**1.2**
X8	3	35.6	**0.8**	3.4	**0.8**	**0.8**	0.8
avg	2.59	26.28	1.36	2.49	1.20	1.29	**1.09**

the azimuth of the navigation course). This is a very competitive score of efficiency, by considering that a new message is processed after an interval of two or more minutes. We also observe that this comparative study of the interpolation time shows different timings depending on the cost used. In particular, when we use H2, which is the azimuth-based cost defined for fitting non-linear courses of navigation, the interpolation time increases. Our interpretation of this result is that this cost, when used alone, frequently prefers to explore curvilinear itineraries. This introduces frequent backtracks into the search. This is confirmed by the fact that H1 or H3, which favor linear motions, exhibit interpolation times which are always lower than those of H2, while the combination of these three costs achieves the lowest interpolation time in general.

The special attention that we devote to H1H2H3 in this analysis depends on the fact that this cost combines all the specific abilities of the cost functions described in Section 3.3 in a single cost, without requiring specific users' decisions. To definitely assess the viability of H1H2H3 in this interpolation task, we also compare accuracy of partial interpolations (sub-itineraries between AIS messages) performed with H1H2H3 and bestH, respectively. bestH denotes the cost function that is the best on average for the vessel (in bold in Table 3) selected per vessel. The pairwise statistical comparison is done with the Wilcoxon Matched-Pairs Signed-Ranks Test. This comparison reveals that H1H2H3 performs, in general, equally well to bestH (column 9 of Table 3). In fact, results show that the hypothesis that bestH and H1H2H3 have identical distributions can be rejected only by resorting to a p-value greater than 0.1. But a p-value of 0.1 is not considered as significant in statistics. This is intended as a point in favor of the generally good performance of H1H2H3 confronted with the best one. This analysis also indicates that, although there is no cost that is the best in absolute, the combination of the several costs produced with H1H2H3 can be preferred to any other cost for its trade-off between accuracy and efficiency.

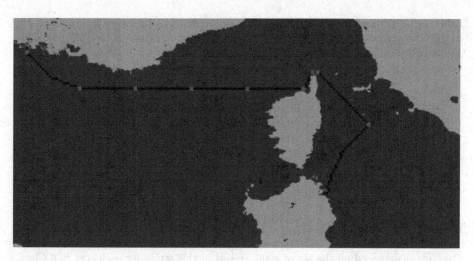

Fig. 3. The vessel itinerary computed by using h()=(H1()+H2()+H3)/3. The red cells geo-locate the AIS messages sent from the monitored vessel to ShipTracker.

AIS Message Sampling. Interpolation results are now investigated when vary-ing the number of AIS messages mined by ShipTracker. This analysis permits us to study the dependence between the accuracy (as well as the time) of inter-polation and the frequency of the interpolated messages. It also shows how the interpolator is robust in the case that the number of interpolated messages per vessel diminishes. This case makes sense in a data warehouse system, where the memory is saved by archiving only a sample of the historical AIS messages of a vessel trip.

Results ($\overline{dtw}$ and $\overline{time}$), collected in Tables 5 and 6, show that by diminishing the number of AIS messages interpolated per vessel, the accuracy of the itinerary interpolated by ShipTracker decreases accordingly. This is expected since we look for the same itinerary, but we use less information (AIS messages) to in-terpolate it. In general, we observe that by halving processed messages, errors double accordingly. On the other hand, the computation time spent on average interpolting the sub-itinerary between each pair of consecutive AIS messages in general grows-up consistently with the length of the interpolated sub-itinerary. There are some exceptions in this trend, but the interpolation time also depends on the location of the sampled messages.

In conclusion, this study contributes to our presentation of ShipTracker both as a system that can operate on-line, by interpolating efficiently messages which can also arrive at rapid rate, and as a system that can robustly operate off-line, by interpolating a reduced number of historic messages stored in a data warehouse.

Table 5. AIS Message Amount (interpolation accuracy): itineraries interpolated by varying the percentage of cells of the real itinerary used to geo-locate an AIS message. (Columns 2-8) The $\overline{dtw}$ is averaged on the set of eight vessels in this study for each cost h().

AIS %	h						
	H1	H2	H3	H1H2	H1H3	H2H3	H1H2H3
10%	**0.45**	0.82	0.62	0.56	0.58	0.56	*0.48*
9%	**0.41**	0.70	0.62	0.54	0.56	0.56	*0.51*
8%	**0.47**	1.01	0.66	0.58	0.58	0.54	*0.48*
7%	**0.52**	1.09	0.79	0.73	0.70	0.67	*0.61*
6%	**0.60**	1.21	0.89	0.76	*0.61*	0.65	0.65
5%	**0.87**	1.48	1.10	0.95	1.03	0.96	*0.90*

Table 6. AIS Messages (interpolation time): itineraries discovered by varying the percentage of cells of the real itinerary used to geo-locate an AIS message. (Columns 2-8) The $\overline{time}$ is averaged on the set of eight vessels in this study for each cost h(). Computation times are obtained by running ten trials of the itinerary discovery algorithm and computing the average time on these trials.

AIS %	h						
	H1	H2	H3	H1H2	H1H3	H2H3	H1H2H3
10%	2.59	26.28	1.36	2.49	1.20	1.29	**1.09**
9%	1.32	32.11	1.93	2.52	1.75	3.78	**1.11**
8%	1.23	45.37	1.21	3.03	3.56	4.86	**1.06**
7%	**1.27**	52.07	4.32	5.18	8.55	6.93	2.73
6%	1.43	163.31	17.32	14.64	1.27	**1.36**	1.55
5%	1.52	144.44	1.99	6.46	**1.49**	1.79	1.98

5   Conclusions

We have presented a novel spatio-temporal data mining system, called Ship-Tracker, which mines the AIS messages as they are sent by a vessel moving in open sea space and uses spatio-temporal knowledge to plausibly interpolate the itinerary followed by the vessel between two consecutive AIS messages.

A graph-based analysis of the task is performed, in order to represent the world map as a graph of land/sea cells. A graph-search algorithm is tailored to address the task of interpolating a (non-linear) itinerary from ubiquitous motion data. The itinerary is a graph path, which connects nodes representing the map cells, where the AIS messages are geo-located. The geographical knowledge which is enclosed both in the map and in the AIS messages (neighboring relations, geographical distance, azimuth) has been considered in the definition of the cost functions, plugged in by ShipTracker, to speed-up the itinerary interpolation over the graph.

The experiments evaluate both the accuracy and efficiency of the proposed system and prove the viability of the itinerary interpolation process, even when we reduce the number of AIS messages processed to complete the interpolation task.

The crucial aspect of the proposed technique is the definition of the cost function. Although experiments suggest some global guidelines to define this function, further work can be done. In particular, we plan to extend the cost function model, in order to account for additional knowledge provided by weather, ocean flow and other vessel itineraries, as well as to investigate mechanisms to weight appropriately the contribution of single costs used in the interpolation process. Additionally, we are extending this study by investigating the algorithms of frequent trajectory discovery from position data [9], in order to be able of mine AIS messages to discover navy routes.

Acknowledgments. This work fulfills the research objectives of the project PRIN 2009 Project "Learning Techniques in Relational Domains and their Applications", funded by the Italian Ministry of University and Research (MIUR). The authors wish to thank Lynn Rudd for her help in reading the manuscript, Guido Colangiuli and Rocco Rana for their support in developing the presented system and running experiments.

References

1. Aronoff, S.: Geographic information systems: a management perspective. WDL Publications, Ottawa (1989)
2. Bertolotto, M., Ray, C., Li, X. (eds.): W2GIS 2008. LNCS, vol. 5373. Springer, Heidelberg (2008)
3. Brown, P., Haas, P.: Techniques for warehousing of sample data. In: Proceedings of the 22nd International Conference on Data Engineering, ICDE 2006, p. 6 (2006)
4. Burrough, P.A.: Principles of geographical information systems for land resources assessment, vol. (12). Clarendon Press (1986)
5. Etienne, L.: Spatio-temporal data mining and classification of ships' trajectories. In: Geomatics Atlantic 2012 (2012)
6. Frank, E., Hall, M., Holmes, G., Kirkby, R., Pfahringer, B., Witten, I.H., Trigg, L.: Weka-a machine learning workbench for data mining. In: Maimon, O., Rokach, L. (eds.) Data Mining and Knowledge Discovery Handbook, pp. 1269–1277. Springer (2010)
7. Hart, P.E., Nilsson, N.J., Raphael, B.: A formal basis for the heuristic determination of minimum cost paths. IEEE Trans. Systems Science and Cybernetics 4(2), 100–107 (1968)
8. Sakoe, H., Chiba, S.: Dynamic programming algorithm optimization for spoken word recognition. In: Readings in Speech Recognition, pp. 159–165. Morgan Kaufmann Publishers Inc. (1990)
9. Schrödl, S., Wagstaff, K., Rogers, S., Langley, P., Wilson, C.: Mining gps traces for map refinement. Data Min. Knowl. Discov. 9(1), 59–87 (2004)
10. Semerdjiev, E., Mihaylova, L., Tzvetan Semerdjiev, V.B.: Interacting multiple model algorithms for manoeuvring ship tracking. In: Multi-Source Data Fusion, Information and Security, vol. 2. ProCon Ltd., Sofia (1999)

11. Sun, L.-H., Shen, J.-H.: Prediction of ship pitching based on support vector machines. In: Proceedings of the 2009 International Conference on Computer Engineering and Technology, ICCET 2009, vol. 1, pp. 379–382. IEEE Computer Society (2009)
12. Zhu, F., Maritime, D.: Mining ship spatial trajectory patterns from ais database for maritime surveillance. In: IEEE International Conference on Emergency Management and Management Sciences (2011)

Social Media as a Source of Sensing to Study City Dynamics and Urban Social Behavior: Approaches, Models, and Opportunities

Thiago H. Silva, Pedro Olmo S. Vaz de Melo, Jussara M. Almeida,
and Antonio A.F. Loureiro

Federal University of Minas Gerais
Department of Computer Science
Belo Horizonte, MG, Brazil
{thiagohs,olmo,jussara,loureiro}@dcc.ufmg.br

Abstract. In order to achieve the concept of ubiquitous computing, popularized by Mark Weiser, is necessary to sense the environment. One alternative is use traditional wireless sensor networks (WSNs). However, WSNs have their limitations, for instance in the sensing of large areas, such as metropolises, because it incurs in high costs to build and maintain such networks. The ubiquity of smart phones associated with the adoption of social media websites, forming what is called participatory sensing systems (PSSs), enables unprecedented opportunities to sense the environment. Particularly, the data sensed by PSSs is very interesting to study city dynamics and urban social behavior. The goal of this work is to survey approaches and models applied to PSSs data aiming the study city dynamics and urban social behavior. Besides that it is also an objective of this work discuss some of the challenges and opportunities when using social media as a source of sensing.

1 Introduction

At the beginning, there were mainframes, shared by a lot of people. Then came the personal computing era, when a person and a machine have a close relationship with each other. Nowadays we are witnessing the beginning of the ubiquitous computing (ubicomp) era, when technology recedes into the background of our lives [1, 2].

Mark Weiser, in his classical article entitled "The computer for the 21st century", that appeared in the Scientific American magazine [3], popularized the concept of ubiquitous computing, which envisions the availability of a computing environment for anyone, anywhere, and at any time. It may involve many wirelessly interconnected devices, not just traditional computers, such as desktops or laptops, but may also include all sorts of objects and entities such as pens, mugs, phones, shoes, and many others.

Although this is not the reality yet, much has been done in this direction in the past 20 years after the publication of Weiser's seminal paper, and the key ingredients are evolving in a favorable direction for it. Observe, for example, the increasing number and popularization of numerous types of portable devices.

A fundamental step to achieve Weiser's vision is to sense the environment. The research in wireless sensor networks (WSNs) has provided several tools, techniques and

M. Atzmueller et al. (Eds.): MUSE/MSM 2012, LNAI 8329, pp. 63–87, 2013.

algorithms to solve the problem of sensing in limited size areas, such as forests or factories [4, 5]. However, traditional WSNs have their limitations, such as the high costs
related to achieve very large coverage spaces, such as metropolises size areas. Consider
the challenges to build and maintain such networks.

In this direction, smart phones are taking center stage as the most widely adopted
and ubiquitous computing device [2]. It is also worth noting that smart phones are increasingly coming with a rich set of embedded sensors, such as GPS, accelerometer,
microphone, camera, gyroscope and digital compass [6].

Social media websites such as Foursquare[1], Instagram[2], Flickr[3], Twitter[4], Waze[5],
and Weddar[6] have started to create new virtual environments that integrates the user interactions and, probably because of that, are becoming very popular. Figure 1 illustrates
the popularity of social media use by showing what happens on the Internet at every
sixty seconds. For instance, we can see that more than 6,600 pictures are uploaded on
Flickr and 320 new accounts and 98,000 tweets are generated on Twitter every minute.
Besides that, Foursquare, created in 2009, registered 5 million users in December 2010,
10 million users in June 2011, and 20 million users in April 2012 [7].

The ubiquity of smartphones, associated with the adoption of social media websites,
enables unprecedented opportunities to study city dynamics and urban social behavior
by analyzing the data generated by users. In this way, we can consider social media as
a source of sensing, each one providing different types of data. In Waze, users report
traffic conditions, in Weddar, users report weather condition. Location sharing services,
such as Foursquare, allows users to share their actual location associated with a specific
category of place (e.g., restaurant). This enables the study of human behavioral patterns,
such as mobility, and also the study of the semantic meaning of places in the city. Social
media systems that allow people connected to the Internet to provide useful information
about the context in which they are inserted at any given moment, as those cited above,
are called participatory sensing systems (PSSs) [8, 9].

Indeed, PSSs have the potential to complement WSNs in many aspects. As WSNs
are typically designed to sense areas of limited size, such as forests and factories, PSSs
can reach areas of varying size and scale, such as large cities, countries or even the entire planet [9]. Furthermore, WSNs are subject to failure, since their operations depend
on proper coordination of actions of their sensor nodes, which have severe hardware
and software restrictions. On the other hand, PSSs are formed by independent and autonomous entities, i.e., humans, which make the task of sensing highly resilient to individual failures. The success of PSSs is directly connected to the popularization of the
smartphones and social media.

The goal of this work is to present the state of the art of the use of participatory
sensing systems to study city dynamics and urban social behavior. It surveys approaches
and models applied to generate context (see Section 2.3 for the definition) from big raw

---

[1] http://www.foursquare.com
[2] http://www.instagram.com
[3] http://www.flickr.com
[4] http://www.twitter.com
[5] http://www.waze.com
[6] http://www.weddar.com

Fig. 1. Things that happen on Internet every sixty seconds. Infographic by - Shanghai Web Designers (http://www.go-globe.com/web-design-shanghai.php).

data obtained by participatory sensing systems. It is worth mentioning that it is not our objective to make an exhaustive survey in this subject. Instead, we discuss a compilation of studies that represent recurrent themes addressed by researchers nowadays. For that, we identified five classes of studies and we named them as: (1) mobility patterns; (2) understanding cities; (3) social patterns; (4) event detection; and (5) human behavior. For each class we highlight the approaches and models applied to create new knowledge and semantic meaning from the big raw data. Besides that we also discuss some of the challenges and opportunities when using social media as a source of sensing.

The rest of this chapter is organized as follows. Section 2 discusses the concept of ubiquitous computing, presenting its definition (Section 2.1), discussing its current state (Section 2.2) and also presenting the concept of context aware computing (Section 2.3), which is a central piece of ubicomp. Section 3 discusses the participation of humans in the sensing process, covering particularities of participatory sensing systems and participatory sensor networks. Section 4 presents the approaches and models used to deal with social media as a sensor, for each one of the five classes of studies considered. Section 5 and Section 6 discuss the challenges and opportunities that emerge when dealing with social media as a source of sensing, respectively. Finally, Section 7 presents the final remarks.

2   Ubiquitous Computing

Modern computing can be divided in three eras. The first is characterized by one single computer (mainframe) owned by an organization and used by many people concurrently. In the second era, a personal computer (PC) is usually owned and used by a single person. In the third era, ubiquitous computing (ubicomp), each person owns and uses many computers, especially small networked portable devices such as smart phones and tablets[1, 2].

Ubiquitous computing is related to mobile computing, although they are not the same thing, neither a superset nor a subset [10] of each other. Mobile computing devices are not mere personal organizers. They are devices (computers with processing power) that contemplate a new paradigm: mobility. Mobility has some constraints, such as finite energy sources. This paradigm is changing the way we work, communicate, have fun, study and do other activities while we are moving [11]. The fact is that ubiquitous computing must support mobility, since motion is an integral part of everyday life. Hence, ubiquitous computing relay on mobile computing, but goes much further.

2.1 Mark Weiser's Visions

To talk about ubiquitous computing we have first to mention Mark Weiser, which has been recognized as the "father" of ubiquitous computing. Weiser, called by many "Visionary", was head of the Computer Science Laboratory at Xerox Palo Alto Research Center (PARC) when he coined the term ubiquitous computing in 1988. When the ubiquitous computing program emerged at PARC, it was at first envisioned only as an answer to what was wrong with personal computing, because they were too complex, too demanding of attention, among others things [12]. During the implementation of the first ubicomp system, Weiser's group realized they were, in fact, starting a post-PC era, in other words, ubicomp was emerging [12].

Mark's vision influenced a countless number of researchers. Almost one quarter of all the papers published in the Ubicomp conference between 2001 and 2005 cite Weiser's foundational articles [13]. Among the Weiser's 'foundational papers' of ubiquitous computing, perhaps the most impacting work is the one entitled "The Computer in the 21st Century", publish in *Scientific American* in 1991. In this paper, Weiser describes the ideal ubicomp future, its purposes, concerns and analogies. To illustrate its ideas he told the story of "Sal", a tale about a single mother and how the world evolves around her needs.

> *"The most profound technologies are those that disappear. They weave themselves into the fabric of everyday life until they are indistinguishable from it"* [3, p. 1].

Weiser believed that the most powerful things are those that are effectively invisible in use. The ideal is to make a computer so embedded, so fitting, so natural, that we use it without even thinking about it. The essence of this vision is making everything easier to do, with fewer mental gymnastics [3, 14].

Second Weiser, the style of computing that has been imposed on users in the first and second modern computing eras (mainframes and PCs, respectively) is too attention consuming, and divorce the users of what is happening around them. In the ubicomp world, as Weiser believed, computation could be integrated with common objects that you might already be using for everyday work practices, rather than considering computation to be a separate activity. If the integration is done well, the envisioned invisibility could be achieved [15, 2].

In order to clarify this concept of invisibility, consider the example based on the familiar printed page (inspired in [2]). To perform a printing it is necessary deposit ink on thin sheets of paper, and a consolidated technology is necessary for that. For a good result it is necessary to ensure that: it must be durable in use; not wick in the paper if wet; among other things. However, we rarely pay attention on the ink technologies when we read printed pages. Instead, we read pages and comprehend ideas, not necessary focusing on the technology, the characteristics of the ink, or the manufacturing process of the paper to be able to use it. In this example, the printing technology got invisible for the user, allowing the higher-level goal of reading a story, or acquiring knowledge. This kind of thinking rarely happens with traditional PCs, which demand the users continuously focus attention on the system, maintaining it and configuring it to complete a task.

Good technology is invisible, staying out of the way of the task, like a good car stays out of the way of driving. Bad technology draws attention to itself, not the task, like a car that needs a tune-up. Computers are mostly not invisible. Ubiquitous computing is about enabling invisibility in computers [16].

2.2   Ubicomp Today

As a promising research area, ubiquitous computing gave us more questions than answers [12], and many of them are still open [15]. There are many ubicomp projects around the world working on ubicomp challenges. Those projects range from prestigious computer science Schools, such as MIT (see several projects from Media Lab[7] for some examples), to mainstream computer companies, such as Microsoft (see the website http://research.microsoft.com/en-us/groups/ubicomp/ with some projects).

Since the early days of ubicomp, one of the main concerns was that computer too often remain the focus of attention, rather than being a tool through which we work, disappearing from our awareness [15]. We may have not achieved the original Weiser's vision about Ubicomp yet. But we can say that the key ingredients are evolving in a favorable direction for it. Many critical items that were rare in early 1900s are now commercially viable. Each year more possibilities for the mainstream application of ubiquitous computing open up.

The future envisioned by Weiser, ubiquitous computing, considers a computing environment in which each person is continually interacting with many wirelessly interconnected devices [15]. Today it is easy to find several microprocessors at home, available, for instance, in alarm clocks, the microwave ovens and in the TV remote controls. They do not qualify as ubicomp devices mainly because they do not communicate with each other, but if we network them together they are an enabling technology for ubicomp [1]. It soon may become a reality. For example, Google has announced in the event Google IO'11[8] an initiative called Android@Home, which allows Android[9] applications to discover, connect and communicate with appliances and devices inside the house. After connecting together several information sources with many information delivery systems we will start to have things, such as, clocks that find out the correct time after a power failure, and microwave ovens that download new recipes.

---

[7] http://www.media.mit.edu
[8] http://www.google.com/events/io/2011
[9] http://code.google.com/android/

Besides that, some of our computing technology are becoming ubiquitous, for in-stance smart phones, which are taking center stage as the most widely adopted and ubiq-uitous computer [2]. When we get used to the possibility of accessing a GPS-connected map, social networks and the Internet anywhere at anytime, we will realize the value of this and it will become part of our lives.

> *"Applications are of course the whole point of ubiquitous computing."* [15, p. 80]

We have to keep in mind that is not just one service that will make computing a dis-appearing technology, but the combination of many. Those services have to be available as needed without extraordinary human intervention [17]. The challenge is to create a new kind of relationship between people and computers, where computers do not de-mand too much attention, letting people live their lives [18]. Application will go beyond the big problems like corporate finance, to the little annoyances such as: where are the car-keys? Can I get a parking place? What is the best route to take now? Which pub should I go in a certain area of the city? [1].

Since ubiquitous computing has intersections with many areas of computing, seve-ral research fields can contribute to its development, including distributed computing, mobile computing, sensor networks, and machine learning. In particular context-aware computing is a key area of research that can help us to meet the original design goals of ubicomp [19].

2.3 Context-Aware Computing

Several context definitions have been proposed. Among them, those presented by Schilit et al. [20], Dey et al. [21], and Pascoe [22] are close to the definition considered by most people as the ideal one. The problem is that those definitions lack for generality. Dey and Abowd [23] proposed the following definition of context:

> *"Context is any information that can be used to characterize the situation of an entity. An entity is a person, place, or object that is considered relevant to the interaction between a user and an application, including the user and applications themselves."* [23]

This is one of the most accepted and accurate definitions currently used by re-searchers. It can be observed that the definition is very general when considering what types of data is context, being wide enough to accept the different needs of each appli-cation. In addition, it is interesting to note that the definition is precise, not requiring a list of specific types or classes of contexts.

In this work we consider social media as a source of sensing. In this case, humans are responsible for sharing data. The data shared by the "sensors" (humans plus his/her portable device) can be then transformed in a context used to study city dynamics and urban social behavior. In the next section we discuss the participation of humans in the sensing process. After that, in Section 4, we discuss the model and approaches used to transform raw data shared by users into context information.

3   Humans in the Sensing Process

3.1   Participatory Sensing Systems

Social media systems that allow people connected to the Internet to provide useful data about the context in which they are at any given moment, such as Waze, Weddar, and Foursquare, are called participatory sensing systems (PSSs). PSS is a concept that originally considers that the shared data is generated automatically, or passively, by sensor readings from portable devices [8]. However, here it is also considered manually, or proactively, user-generated data. Systems with those characteristics have been called ubiquitous crowdsourcing [24]. The popularity of participatory sensing systems grew rapidly with the widespread adoption of sensor-embedded and Internet-enabled cell phones. These devices have become a powerful platform that encompasses sensing, computing and communication capabilities, being able to generate both manual and pre-programmed data.

To illustrate this type of system, consider an application for transit monitoring, like Waze. Users can share observations about accidents or potholes manually. Additionally, an application can calculate and share automatically a car speed based on GPS data. Since in this specific case users operate an application that was designed for a specific purpose, the sensed data is structured. If, instead, users use a microblog (e.g., Twitter), the sensed data would be unstructured. For instance, user "Smith" sends the message "traffic now is too slow near the main entrance of the university campus".

3.2   Participatory Sensor Networks

Participatory sensor networks have their users with their portable devices as the fundamental building block. Individuals carrying these devices are able to sense the environment and to make relevant observations at a personal level. Thus, each node in a participatory sensor network consists of the user plus his/her mobile device, with the goal of sending data to the systems. After that, the data usually can be collected throughout services APIs.

Similar to WSNs, the sensed data is sent to the server, or "sink node". But unlike WSNs, PSNs have the following characteristics: (i) nodes are autonomous mobile entities, i.e., a person with a mobile device; (ii) the cost of the network is distributed among the nodes, providing a global scale; (iii) sensing depends on the willingness of people to participate in the sensing process; (iv) nodes transmit the sensed data directly to the sink; (v) nodes do not suffer from severe power limitations; and (vi) the sink node only receives data and does not have direct control over the nodes. More details about PSNs can be found in [9, 25].

Figure 2 shows the components of a participatory sensor network. In particular, this figure highlights the three most important components, namely)(i) the social media as a source of sensing, (ii) the big raw data, and (iii) the context information.

The component "Social media as a sensor" encompasses users sharing data through social media systems. The component "Big raw data" is responsible for data management. As we can see in the Figure 2, the collection process may be repeated, for example, to get redundant or complementary data from other social media systems.

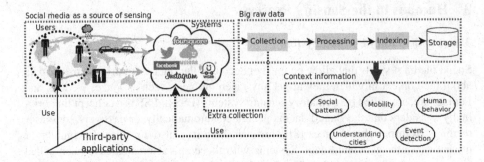

Fig. 2. Overview of a participatory sensor network

After that, the collected data needs to be processed in order to be stored. Since the amount of data coming from participatory sensing systems may be very large, all the components need to be carefully designed if the goal is to get real-time information. A more detailed discussion of some of the challenges is presented in Section 5.

After the data management stage, the data is ready to be analyzed. The component "Context information" represents five type of analysis that could be performed: Social patterns; mobility; understanding cities; human behavior; and event detection. All these classes of analysis are discussed on Section 4.

4   Approaches Used to Deal with Social Media as a Sensor

In this section we discuss the approaches and models used to extract and generate context information from participatory sensing systems data in order to study city dynamics and urban social behavior. This section will concentrate in the component named "Context information", shown in Figure 2. In this particular component is represented different classes of studies, and they will be discussed here. Section 4.1 discusses studies related to the analysis of mobility patterns. Section 4.2 considers studies that focus on the better understanding of city dynamics. Section 4.3 discusses the study of social patterns. Section 4.4 discusses studies concerned in event detection. And finally, Section 4.5 presents studies related to human behavior.

It is worth mentioning that each class of study is not necessarily mutually exclusive. For example, Long et al. [26] used a Foursquare dataset to classify venues based on users' trajectories. This work has intersections with the class "Mobility patterns" (Section 4.1), but instead of being classified in that class, it was assigned to the class "Understanding cities" (Section 4.2), since it is more concerned in the analysis of city dynamics. The main goal of this section is to present hot research topics, and present what have been done to address some of the challenges.

4.1   Mobility Patterns

This class of work focuses on studying mobility patterns of users from their logs generated from social media websites. These logs usually include spatio-temporal information, e.g., check-ins and geolocated photos. The study of mobility is useful for many purposes. For example, it is possible to understand how human allocate time to different activities, thus being a fundamental and traditional question in social science [27]. As another example, one could design new tools to help traffic engineers to understand the flow of people.

The modeling of mobility patterns has been attracting the attention of researchers in different fields, such as physics and ubiquitous computing [28–30]. For example, Song et al. [31] analyzed 50,000 cellphone users and showed that user mobility presents high predictability. It is important to point out that data derived from social media is different from GPS tracking or cellphone usage data, such as phone calls, and present special features and varied contexts. For example, check-ins in location sharing services or photos shared in a photo sharing service bring extra information of a particular place. For instance, a check-in is associated with a type of venue, e.g. pub, and a photo may bring the information about the current situation inside this venue. Again, throughout this work our focus is on studies that analyze data from social media.

Cheng et al. [32] analyzed 22 million check-ins posted from several location sharing services (Foursquare is responsible for 53.5% of the total). They found that users follow simple and reproducible patterns, and also that social status, in addition to geographic and economic factors, are coupled with mobility. **Approach:** to make their analysis they used three statistical properties to study and model human mobility patterns: *displacement*; *radius of gyration*; and *returning probability*. The *displacement* of check-ins is the distance between consecutive check-ins, measuring how far a user has moved. The *radius of gyration* measures the standard deviation of distances between the users' check-ins and the users' center mass. This measure indicates how far and frequently a user has traveled. *Returning probability* is a measure of periodic behavior in human mobility, since periodic behavior tends to happen frequently due to human routines. Besides that, the authors also studied factors that could influence mobility, such as social status and geographic and economic constraints.

Cho et al. [33] investigated patterns of human mobility in three datasets: check-ins in two location sharing services and cellphone location data. They were particularly interested in determining how often users move and where they go to, as well as how social ties may impact their movements. They observed that short-ranged travel is periodic both spatially and temporally and is not affected by the social network structure, while long-distance travel is more inuenced by social network ties. **Approach:** based on their empirical findings they built a model named Periodic & Social Mobility Model to predict mobility of users. This model is composed by three parts: (1) a model of spatial locations that a user usually visits based in a two-state mixture of Gaussians with a time-dependent state prior; (2) a model of temporal movement between these locations based on a truncated Gaussian distribution parameterized by the time of the day; (3) a model of movement that is influenced by the ties of the social network, e.g. encountering friends. In this specific model, if a user performs a check-in, then it will more likely be close in space and time to one of his/her friend's check-ins. Their model is able to predict the exact user location at any time with 40% accuracy.

Nguyen and Szymanski [34] used Gowalla, a location-based social network, to create and validate models of human mobility and relationships. In that work, the authors proposed a friendship-based mobility model (FMM) that take into account social links in order to provide a more accurate and complex model of human mobility. With this model the authors were able to study how frequently friends travel together. This model may improve the accuracy of a varied number of applications, such as traffic engineering in communication networks, transportation systems, and urban planning. **Approach:** the proposed mobility model uses a Markov Model where the states represent locations of check-ins and the links represent the probability of going from one place to another. For example, the probability of going from work to pub is defined as the ratio between the number of times a given user performs a check-in in a pub right after a check-in at work, and the number of times that user performs a check-in at work.

Zheng et al. [35] studied tourist mobility and travel patterns from geotagged photos shared on Flickr. In order to extract the travel patterns, the authors focused the analysis on tourist movement according to regions of attraction and topological characteristics of travel routes by different tourists. The authors demonstrated its potential by testing the approach on four cities. **Approach:** first it is built a database of touristic travel paths based on the concept of mobility entropy (considering Shannon's entropy), used to discriminate the touristic and non-touristic movement. Then, a significance test is applied to ensure that the resulting path is statistically reliable. For that, they devised two methods, one based on a Poisson distribution and the other on a normal distribution. Next, it is proposed a method to discover regions of attraction in a city, using for this the DBSCAN clustering algorithm. To study the touristic movement the authors considered a Markov chain model created from the visiting sequence of regions of attraction discovered by the proposed method. With that, they can estimate statistics of visitors traveling from one region to another. In order to study the topological characteristics of tour routes, the authors perform sequence clustering on travel routes, applying a modified version of the longest common subsequence as a similarity metric to minimize noise.

4.2 Understanding Cities

Information obtained from participatory sensing systems have the power to change our perceived physical boundaries and notions of space [36], as well as to understand city dynamics better. This section focuses in presenting studies in these directions. Many potential applications can benefit from these types of studies, such as tools for city planners to provide new manners to see the city, or for end users who are looking for new ways to explore the city.

Cranshaw et al. [37] presented a model to extract distinct regions of a city that reflect current collective activity patterns. The idea is to expose the dynamic nature of local urban areas considering spatial proximity (derived from geographic coordinates) and social proximity (derived from the distribution of check-ins) of venues. **Approach:** in their study the authors considered data from Foursquare. In order to explore this data, the authors developed a model based on spectral clustering. One of the main contributions is the design of an affinity matrix between venues that effectively blends spatial proximity and social proximity. The similarity of venues is then obtained by comparing pairs of these dimensions. After that, this is used to compute the clusters that may represent different geographical boundaries of neighborhoods. The clustering method is a variation of the spectral clustering proposed by Ng et al. [38], introducing a post processing step to clean up any degenerated cluster.

Noulas et al. [39], proposed an approach to classify areas and users of a city by using venues' categories of Foursquare. This could be used to identify users' communities that visit similar categories of places, useful to recommendation systems, or in the comparison of urban areas within and across cities. **Approach:** their approach is based on spectral clustering algorithm [40, 38]. More specifically, the authors divide the area of a city to be analyzed into a number of equally sized squares, each of them will be a datapoint input for the clustering algorithm. For each area it is represented the activity performed on it based on check-ins in each existing category on that area. Then, it is calculated the similarity between two areas as the cosine similarity between their corresponding activity representation. Having the similarity information, the authors apply it in the spectral clustering algorithm.

Long et al. [26] used a Foursquare dataset to classify venues based on users' trajectories. The premise is that the venues that appear together in many users trajectories will probably be taken as geographic topics, for example representing restaurants people usually go to after shopping at a mall. The approach can be applied, for instance, to understand users' preferences to make recommendation of venues. **Approach:** the authors used the Latent Dirichlet Allocation (LDA) [41] model to discover the local geographic topics from the check-ins. With this approach, it is possible to dynamically categorize the venues in Foursquare according to the users' trajectories, what indicates the crowds preferences of venues. LDA is usually used to cluster documents based on the topics contained in a corpus of documents. For this reason, some terms used to describe the modeling make reference to this context. The authors considered that a single check-in represents a word, which is the basic unit in the LDA. A trajectory of a user consists of all the venues visited by him/her, and this represents a document in the analogy.

In our previous work [42], we propose a technique called City Image and we show its applicability using eight different cities as examples. The resulting image is a way of summarizing the city dynamics based on transition graphs, which map the movements of individuals in a PSN. This technique is a promising way to better understand the city dynamics, helping us to visualize the common routines of their citizens.. **Approach:** the proposed technique consists of a square matrix that summarizes the city dynamics. This matrix is constructed from two transition graphs. First, we construct a transition graph $G(V, E)$, where the nodes $v_i \in V$ are the main categories of the locations and an edge (i, j) exists from node v_i to node v_j if at some point in time an individual performed a check-in in a location categorized by v_j just after performing a check-in in a location categorized by v_i. The weight $w(i, j)$ of an edge is the total number of transitions that occurred from node v_i to node v_j. After constructing G, we create ten random graphs $G_{Ri}(V, E_{Ri})$, where $i = 1, \ldots, 10$ and each one is constructed using the same number of transitions used to construct G. When constructing this random graph instead of considering the actual transition $v_i \rightarrow v_j$ performed by an individual, e.g., "Smith", we randomly pick a location category to replace v_j, simulating, then, a random walk for this individual. We use the distribution for the randomly generated edge weight values for $G_{R1..10}$ to build three ranges: *rejection range* (representing transitions that are not likely to happen), *favouring range* (representing transitions that are likely to happen), *indifference range* (representing transitions that neutral to be performed by users). These ranges are expressed in the visualization.

Kisilevich et al. [43] used geo-tagged photos obtained from Flickr to analyze and compare temporal events that happened in a city, and also to rank sightseeing places. More specifically, the authors presented a way to assess the attractiveness of places based on their positions in a ranking, and suggested a set of visual analytic methods that mixes computational techniques with visual interactivity in order to support analysis of the data. **Approach:** to find the attractiveness of places the authors applied the algorithm DBSCAN [44]. In order to highlight areas of people's activities within a cluster, the authors applied density maps. From the clusters obtained in the clustering part, the weight of every geotagged photo is calculated using a density function based on the relative position of photos of other users in a cluster. The calculated weight is mapped to a color, facilitating the visual inspection.

Frias-Martinez et al. [45] used a dataset from Twitter and proposed a technique to determine the type of activities that is most common in a city by studying tweeting patterns. They also proposed another technique to automatically identify landmarks in a city. **Approach:** to automatically identity urban land usage, the authors apply two methods. The first one is land segmentation. For that it is applied Self-Organizing Maps [46], which is an unsupervised neural network. After training the network, it is obtained a map that segments the urban land into geographical areas with different concentrations of tweets. Each neuron of the network represents a pointer to a region with a high density of tweets. With that, the authors apply Voronoi tessellation considering the location of the neurons to compute the land segments. Next, the authors use the segments found to detect different land usages considering the average tweet usage on them. So, for each land segment is built a unique tweet-activity vector that represents the average tweeting temporal behavior. To characterize urban land usage, it is applied the k-means algorithm, which shows common tweeting behavior across land segments. To identify the landmarks, the authors used the mean-shift clustering technique [47]. The authors considered in this algorithm that every tweet is assigned to a local maxima and a cluster represents a potential landmark. After the execution, if the resulting clusters are ranked by the number of tweets on them, then the result represents a list of the most popular landmarks.

Ji et al. [48] mine blog-based sight photos in order to discover and summarize city landmarks. Their main contribution is a generalized graph modeling framework. This study is useful, for example, for personalized tourist suggestions. **Approach:** first the authors have to extract locations of photos. For that, they collect photos with different descriptors. To identify their locations they use an application called gazetteer [49], which is able to identify location from web resources. Then they create a hierarchical visual-textual clustering scheme to organize sight photos into a "scene-view" structure for each city. For this purpose it is used the concept Bag-of-Visual-Words [50] to generate the content descriptor of photos. Bag-of-Visual-Words are clustered by their similarity measured by the cosine distance, generating then "views". After that the authors create a "scene-view", using textual clustering to aggregate "views" into "scenes". Next, they model two different graphs. The first one represents a scene, where each node is a photo and an edge exist if there is at least one word identical in the photos descriptors. For this graph they present an algorithm, PhotoRank, to discover representative views within a scene. Finally, the authors create another graph to represent the city, that encompasses a scene layer, and present an algorithm to discover city landmarks on it, which explores the PhotoRank algorithm and is inspired in [51].

4.3 Social Patterns

This class of studies concentrates in the analysis of data from social media to understand social patterns. Data from social media enables unprecedented opportunities to study human relationships in a global scale, at a relatively low cost. Examples of possibilities include community detection, products recommendation based on the discovery of similar socio-economic behavior, and new definitions of network centrality.

Scellato et al. [52] presents a study of the spatial properties of the location-based social networks arising among users. Among the results, the authors reported, for instance, that 40% of social links happens below 100 km, and that there is strong heterogeneity across users related to both social and spatial factors. **Approach:** to extract properties and verify their hypothesis, the authors analyzed datasets of three location based services: Foursquare, Gowalla, and Brightkite. In their study, the authors used two randomized models, a social model and a spatial/geo model, to assess the statistical signicance of the empirical spatial properties of the networks analyzed. The social model keeps the social connections as they are, randomizing all user locations. The geo model keeps the user locations unmodied and then assigns every social link between two users at a certain distance according to the relative probability of friendship, observed in their analysis.

Cranshaw et al. [53] introduced a new set of features of human location trail for analyzing the social context of a geographical region. They demonstrated the applicability of these features by presenting a model for predicting friendship between two users, showing significant gains over previous models for the same purpose. **Approach:** the authors used a dataset from Locaccino[10], a system that allows users to share his/her current location with other Locaccino users through Facebook[11]. For the co-location analysis, the authors split the space in grids of 0.0002 x 0.0002 latitude/longitude, which means approximately 30 meters x 30 meters. The time was considered in slots of 10 minutes. In this way, a user is co-located with another user if they are located in the same grid within a slot of time. To model the co-location of users, it is applied three diversity measures: frequency, user count, and entropy (Shannon's entropy). The frequency measure captures the raw count of users who visit a location. The user measure considers the total number of unique users in a location. The entropy measure considers the number of users observed at the location, as well as the relative proportions of observations. High entropy means that many users were observed at the location with equal proportion.

Quercia et al. [54] study how social media communities resemble real-life ones. They tested whether established sociological theories of real-life social networks still hold in Twitter. They found, for example, that social brokers in Twitter are opinion

---

[10] http://www.locaccino.org
[11] http://www.facebook.com

leaders who take the risk of tweeting about different topics. They also discovered that most users have geographically local networks, and that social brokers express not only positive but also negative emotions. **Approach:** the authors applied network metrics about topic, geography, and emotions, regarding to parts of one's social world. These metrics include reciprocity, simmelian ties, and network constraint. Reciprocity is the proportion of edges in a network that are bidirectional. Simmelian ties are a measure that considers triadic relationships. Network constraint measure brokerage opportunities in the network, where high network constraint means less brokerage opportunities. They used Burt's formulation [55] in this specific case.

Java et al. [56] studied blog communities. For that they present a technique for clustering communities by using both the hyperlink structure of blog articles and tag information available on them. The technique was tested in a real network of blogs and tag information, as well as in a citation network. **Approach:** the authors define a community as a set of nodes in a graph that link more frequently within this set than outside it, and they also share similar tags. Their technique is based on the Normalized Cut (NCut) algorithm [57].

Sadilek et al. [58] studied the interplay between people's location, interactions, and their social ties, presenting a technique for inferring link and location information from a stream of message updates. The authors demonstrated, by analyzing users from New York City and Los Angeles, that their technique significantly outperforms other current comparable approaches. **Approach:** for link prediction their approach infers social ties by considering patterns in friendship formation, the content of people's messages, and user location. For location prediction, their technique implements a probabilistic model of human mobility, where it treats users with known GPS positions as noisy sensors of the location of their friends.

4.4   Event Detection

This class of work is focused in the identification of events through data shared in social media. This task is especially favorable due the real-time nature of certain types of social media, such as Twitter. Events might be natural ones, such as earthquakes, or not natural ones, such as the identification/prediction of stock market changes.

Bollen et al. [59] studied whether collective mood states derived from Twitter feeds are correlated to the value of the Down Jones Industrial Average (DJIA) over time. Their findings indicate that it is possible to obtain an accuracy of 86.7% in predicting the daily up and down changes in the closing values events of the DJIA. This is possible by choosing specific mood dimensions, but not all that were considered. **Approach:** to extract the sentiment expressed by the users in the tweet the authors used two tools. The first one is the OpinionFinder (OF), which extract negative or positive sentiments from the message. The second tool, Google-Profile Mood State (GPOM), extract six-dimensional daily time series of public mood. The authors use Granger causality analysis in which it is correlates DJIA values to GPOMs and OF values of n past days. The authors also trained a Self-Organizing Fuzzy Neural Network to predict DJIA values on the basis of various combinations of past DJIA values and OF and GPOMS public mood data.

Sakaki et al. [60] studied the real-time interaction of events in Twitter (e.g. earth-quakes), and propose an algorithm to monitor tweets to detect a target event. To de-monstrate the effectiveness of their method, the authors built an earthquake reporting system in Japan, which was capable to detect 96% of earthquakes reported by the Japan Meteorological Agency (JMA) with seismic intensity scale of 3 or more. Notification to registered users was delivered faster than the announcements that are broadcast by the JMA. **Approach:** the authors devise a classifier of tweets based on features such as the keywords in a tweet, the number of words, and their context. After that, they produced a probabilistic spatio-temporal model for the target event that can find the center and the trajectory of the event location.

Lee and Sumiya [61] present a geo-social event detection system to identity local events (e.g., local festivals) by monitoring crowd behaviors indirectly via Twitter. The system was created on the hypothesis that users probably write many posts about these local events. **Approach:** first the authors decide what the usual status of crowd behav-iors is in a geographical region in terms of tweeting patterns. After that, a sudden in-crease in tweets in a geographical region can be an important clue. Another hint might be the increasing number of Twitter users in a geographical region in a short period of time. The authors also consider if the movements of the local users become un-expectedly elevated. The detection of unusual events in the study uses the concept of boxplot [62], which is applied to create ranges to determine the cases desired to be detected.

Becker et al. [63] analyze streams of Twitter messages to distinguish between mes-sages about real-world events and non-event messages. They identify each event and its associated Twitter messages. **Approach:** the authors use an online clustering technique that groups together similar tweets. With that, they extract features for each cluster to help determine which clusters correspond to events. Next, the authors use these features to train a classier to distinguish between event and non-event clusters.

4.5 Human Behavior

This group of studies focus on the study of human behavior through the data shared in social media, which, as we mentioned before, can be seen as signals given by users. This type of study can be applied, for example, to the discovery of individual social proles, the discovery of collective behaviors, the analysis of sentiment and opinion evolution, and a better understanding of why individuals take certain actions.

Joseph et al. [64] analyzed a Foursquare dataset to identify groups of people and the places they go. Their model was able to identify groups of people which represent both geo-spatially close groups and people who appear to have similar interests. **Approach:** their model is based on the idea of topic modeling. For that they applied the Latent Dirichlet Allocation [41]. In the model instantiation, each check-in for a user can be thought of as a word in a document. Similar to text documents, where a "document" can have multiple words, the authors dened a multinomial distribution for the check-ins for each user by using the number of check-ins in each venue as features.

Naaman et al. [65] focused their study in the characterization of tweeting patterns in different cities located in the USA, envisioning to provide a framework for reasoning about activities performed in cities. This study might be useful to deal with challenges such as transportation or resource planning faced in urban studies. **Approach:** first the authors selected tweets from some US cities. Then, they selected the top 1000 words from the resulting dataset, and made a cleaning procedure in this dataset using the NLTK toolkit[12], removing, for example, stopwords. After that, the authors performed a study of keyword-based diurnal patterns in the considered locations. Besides that, the authors applied the concept of Shannon's entropy and Mean Absolute Percentage Error (MAPE), to measure the variability of the data within days and across days, respectively.

Poblete et al. [66] analyzed a twitter dataset aiming the discovery of insights of how tweeting behavior varies across countries, as well as the possible explanations for these differences. **Approach:** first the authors selected the top ten most active countries. Then, they extracted differences in the number of twittes per user, languages used per country, sentiment analysis (happiness), using the Affective Norms for English Words (ANEW) [67] and a Spanish version of it [68], and the content of the tweet. Moreover, they studied the social network properties for each country applying metrics, such as, clustering coefficient, diameter, and shortest paths.

Gao et al. [69] propose a model to address the "cold start" location prediction problem, by using the social network information. Results in an experiment based on a real-world location-based social network show that the approach is effective for the studied problem. **Approach:** the authors' strategy encompasses the investigation of the check-ins behavior to understand the correlations in the context of the user's social network and geographical distance. For this analysis, they considered four social cycles. With that, the authors modeled the geo-social correlations of "new check-in" behavior considering the intrinsic patterns of users' check-ins and his/her social cycles.

Yu et al. [70] used the users' behavioral patterns extracted from a Sina Weibo[13] to investigate how users' frequent activities reflect their sleeping time and living time zones. The authors showed that may be possible to detect the sleeping time of users. Their results could also be used as an alternative way to estimate time zones. **Approach:** based on the time series of the Sina Weibo usage the authors applied a simple statistical method, assuming that users keep a daily routine, going to bed and waking up on time, to detect long periods of inactivity.

5   Challenges

Considering social media as a source of sensing, constructing then a participatory sensor network imposes many challenges. Looking at Figure 2 we see that a participatory sensor network could be divided in different blocks. In Section 4 we described how researchers have been addressing challenges mainly related in the block named "Context information", which represents models and approaches to transform big raw data from social media in useful information, to be applied, for example, in applications. In this section we are concentrated in challenges related with the blocks named "Social media as a source of sensing" and "Big raw data".

Among the challenges present in these blocks we can mention data quality, data collection, data storage, data processing and indexing. The quality of the shared data is a

---

[12] http://www.nltk.org

[13] A popular Chinese micro-blogging service.

challenge that has been relatively well tackled in the web domain, however there are unique challenges for controlling the quality of shared data when dealing with ubiquitous user contributions [24]. For instance, since users can produce sensor readings with relatively little effort, data integrity is not always guaranteed [71].

Besides that the shared data through social media in some cases is free text, not presenting structure nor codified semantics, being complex to understand and process. To better interpret such complex data, visualization techniques and tools should be developed. Another issue related to data quality is the liberty given to users in certain social media systems. Sometimes, users can post whatever, even incorrect, information in different formats. This demands mechanisms for data filtering. A reputation system may be very useful in this case.

Data collection is a challenging issue especially from third-party social media services, such as Foursquare and Waze. By default, data shared in those systems are usually private, unless users decide to make them public, for example sharing it on Twitter. This means that no public data can be available at all. Furthermore, since the data depends on the users will in contribute, there is no guarantee on the delivery of any data. This makes the use of social media as a source of sensing completely out of the control loop of system managers and application developers. Some actions can be taken to ensure that the user participation is sustained over time. An example of action could be an incentive mechanism based on micro-payments, i.e., every time a user perform a given activity, he/she receives an small payment, as proposed by Reddy et al. [72].

Another important issue is deal with a huge volume of data that social media systems can offer, because it tend to be large and complex being difficult to process and index using traditional database management tools or data processing applications. This imposes challenging issues to offer real-time services using a participatory sensor network. To tackle this issue we need methods to effectively store, move and process big amounts of data. New algorithmic paradigms, for example map-reduce, should be designed, as well as specific mining techniques should be created according to these new paradigms. Other methods should contemplate data engineering approaches for large networks with up to billions nodes/edges, including effective compression, search, and pattern matching methods [27].

Furthermore, participatory sensor networks are very dynamic. To illustrate the challenges that emerge with this characteristic we analyze the information flow in PSNs, which is depicted in Figure 2, particularly the two flows symbolized by the arrows labeled with the word "use", directing from Context information component to Systems, and from Third-party applications to Users. Users rely on applications, such as Twitter or Waze, to transmit their sensed data. The sensed data is, then, transmitted to the server, or the "sink node". The Context information component is responsible for processing the shared data and generating useful information, or contexts (Section 2.3). The systems, such as Waze, by its turn, may be fed back with the generated contexts and, from this, provides useful information to the users. Contexts can also be generated by third-party applications. For example, in Section 4, we describe an example of application that enables the identification of regions of interest in a city, which exemplifies a type context. After using this application, users may choose to change their behavior, e.g., to visit preferably popular areas, which may ultimately impact the number of potential shared data in those places. This gives an idea of how dynamic a participatory sensor network is and the challenges that emerge to deal with this dynamism.

Besides those problems there is still the problem of user's privacy. This challenge is very broad, being present in many layers of the system. Data privacy in social media systems has been currently discussed in several studies, such as: [73–75].

A wide range of novel applications opens up after dealing with the challenges of this research field. Some of the opportunities are illustrated in the next section.

6   Opportunities

In this section we present some of the promising opportunities when considering social media for the large scale study of city dynamics and urban social behavior. For that we use the Foursquare dataset analyzed in the study [42].

Nodes in a participatory sensor network move according to their routines or local preferences, and this is interesting for applications that want to capture city dynamics and urban social behavior. In this direction we present several opportunities, grouped in two categories: *Area semantics* (Section 6.1); and *Urban transitions* (Section 6.2).

6.1   Area Semantics

There are many opportunities to design semantic location services, and this sort of services will be crucial for the next wave of killer applications [76]. The opportunities pointed here exploit the information about the category of the venues present in the considered dataset. A complete list of these categories, with examples, can be found in [42].

Application accessed mostly by smartphones provides datasets that represent the social network topology and dynamics of entire cities, enabling the analysis of the social, economic, and cultural aspects of particular areas. For example, regions that provide a small amount of data compared to other regions of the same city might indicate a lack of technology access by the population, since the frequent use of location sharing services often relies on smartphones and 3G or 4G data plans that are expensive in some countries. The preliminary results in the use of participatory sensor networks in these scenarios demonstrate good opportunities to enable the visualization of interesting facts. For example, analyzing carefully the data for the particular case of Rio de Janeiro, shown in Figure 3, we observe that it is common to find very poor areas next to wealthy ones. Note the small sensing activity in the circle areas indicated as poor areas. This information may be useful, for example, to guide better public politics in those areas. The same information can be obtained using traditional methods, such as surveys, but in this new way we may be able to obtain them in an automatic and cheaper way using a participatory sensor network. For this purpose, similar algorithms to the one proposed in [37] could be applied.

Other opportunities to classify areas emerge when jointly considering the time and venue where the check-ins are performed. It may be possible to visualize crowds in a city in near real-time. Besides that, humans have seasonal patterns due to their routines. This seasonality has a great potential for prediction applications, since it is very likely

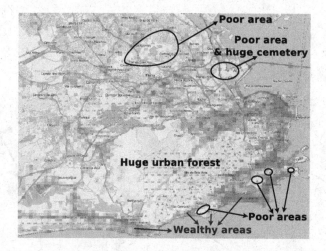

Fig. 3. Example of possible area classification by lack of sensing

that people periodically repeat their activities. We do believe that there are many opportunities for prediction given by the circadian rhythm of people, enabling the prediction, for example, of crowds. This type of information is valuable in many scenarios, such as services for smart cities to avoid traffic in certain areas and offer alternative routes for users. For instance, Hsieh et al. [77] proposed a time-sensitive model to recommend trip routes based on the information extracted from Gowalla check-ins.

In one of our previous study [78] we present a scenario that illustrates another opportunity that exploits the same data. For that, we created a simple method to estimate the number of check-ins in certain time and space, taking into account different categories of places. We show that temporally it is possible to distinguish popular areas in different regions of the city, and this might be useful as on decision criterion when choosing an area to visit at a certain time.

6.2   Urban Transitions

Now we present another range of opportunities that emerges from urban transition graphs. The urban transition graph maps the movements of users between locations. This graph is a directed weighted graph $G(V, E)$, where a node $v_i \in V$ is a specific location (e.g., Times Square) and a direct edge $(i, j) \in E$ marks a transition between locations. That is, an edge exists from node v_i to node v_j if at least one user performed a check-in in the location represented by v_j just after performing a check-in in the location represented by v_i. The weight $w(i, j)$ of an edge is the total number of transitions that occurred from v_i to v_j.

Here we consider the same requirements for transitions specified in [78]. Figure 4 shows heavy weighted edges and hub nodes (top 50 edge weights and node degrees) for Belo Horizonte, Mexico City, New York and Tokyo. Stars represent the hubs, black arrows represent the edges, and black circles represent self-loops. The larger[14] the

---

[14] Numbers grow in logarithmic scale.

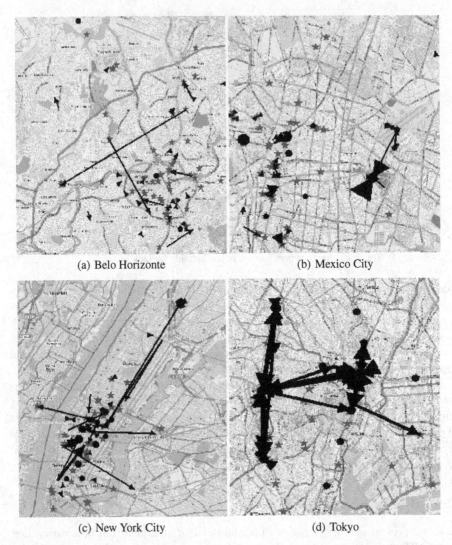

(a) Belo Horizonte   (b) Mexico City

(c) New York City   (d) Tokyo

Fig. 4. Top 50 edge weights and node degrees (hubs) for 4 cities. Stars represent hubs, black arrows edges, and black circles self-loops. The larger the symbol, the higher the value.

symbol, the larger the value. Note that the city flow is very concentrated and skewed, as expected, with a small fraction of the city areas having most of the heavy weighted edges and hubs. Note also for Belo Horizonte and Mexico City that most of the heavy weighted edges are self-loops and low distance edges, implying that people tend to perform activities in the neighborhood of where they are while they can. On the other hand, for New York City and Tokyo, cities that are known for their fast public transportation systems, favor the existence of some long distance heavy weighted edges along the public transport links.

This scheme may be used to support various applications, for example, a public information dissemination scheme, which usually shows traditional advertising. If one knows where the city hubs are, he/she could strategically put these displays in these locations. Moreover, if one verifies an unusual and constant flow of people between two independent business venues in a city, the owners could sign a commercial agreement to increase their revenues by, for example, advertising each other's businesses.

Urban transition graphs are useful also to display a visualization of a city based on the transitions that are likely to occur, as the demonstrated by the City Image technique presented in our previous work [42]. The city image can be expanded to consider sub-categories instead of main categories. Since PSN data is highly skewed, few top transitions between sub-categories should be good indicators of the city dynamics. This technique could be useful as a way to measure the distance between two cities, enabling cities comparison and clustering worldwide that could be interesting for recommendation systems.

Other example in this direction is the study performed by Long et al. [26], which used a Foursquare dataset to classify venues based on users' trajectories. The study performed by Zheng et al. [35] is also one more example, since the authors show the potential in exploring transitions from geotagged photos shared on Flickr.

In this section our goal was illustrate some of the open opportunities in this field. Certainly there are many others.

7 Final Remarks

In this chapter we present the state of the art of the use of participatory sensing systems to study city dynamics and urban social behavior. This work surveys approaches and models applied to generate context from raw data obtained from PSSs. To achieve this goal we studied a compilation of studies that represent five recurrent themes addressed by researchers nowadays, namely: (1) mobility patterns; (2) understanding cities; (3) social patterns; (4) event detection; and (5) human behavior. For each class we highlight, for each study, the approaches and models applied to create new knowledge and semantic meaning from the big raw data. Besides that we also demonstrate a range of fruitful opportunities that emerge when using participatory sensing to the large scale study of city dynamics and urban social behavior.

Acknowledgment. This work is partially supported by the INCT-Web (MCT/CNPq grant 57.3871/2008-6), and by the authors individual grants and scholarships from CNPq, CAPES (scholarship 7356/12-9), and FAPEMIG.

References

1. Weiser, M., Brown, J.S.: The coming age of calm technology (October 1996)
2. Krumm, J.: Ubiquitous Computing Fundamentals, 1st edn. Chapman & Hall/CRC (2009)
3. Weiser, M.: The Computer in the 21st Century. Scientific American 265(3), 94–104 (1991)
4. Yick, J., Mukherjee, B., Ghosal, D.: Wireless sensor network survey. Computer Networks 52(12), 2292–2330 (2008)

5. Akyildiz, I., Su, W., Sankarasubramaniam, Y., Cayirci, E.: Wireless sensor networks: a survey. Computer Networks 38(4), 393–422 (2002)
6. Lane, N., Miluzzo, E., Lu, H., Peebles, D., Choudhury, T., Campbell, A.: A Survey of Mobile Phone Sensing. IEEE Communications Magazine 48(9), 140–150 (2010)
7. Molina, B.: Foursquare tops 20 million users. USA Today (April 2012),
   http://content.usatoday.com/communities/technologylive/
   post/2012/04/foursquare-tops-20-million-users/1
8. Burke, J., Estrin, D., Hansen, M., Parker, A., Ramanathan, N., Reddy, S., Srivastava, M.B.: Participatory sensing. In: Workshop on World-Sensor-Web (WSW 2006): Mobile Device Centric Sensor Networks and Applications, pp. 117–134 (2006)
9. Silva, T.H., Vaz de Melo, P.O.S., Almeida, J.M., Loureiro, A.A.F.: Uncovering Properties in Participatory Sensor Networks. In: Proc. of the 4th ACM International Workshop on Hot Topics in Planet-scale Measurement (HotPlanet 2012) (June 2012)
10. Weiser, M.: Weiser's website about ubicomp (1996),
    http://www.ubiq.com/weiser/weiser.html (Website accesed for the last time in May of 2013)
11. Satyanarayanan, M.: Fundamental challenges in mobile computing. In: Proceedings of the Fifteenth Annual ACM Symposium on Principles of Distributed Computing, PODC 1996, pp. 1–7. ACM, Philadelphia (1996)
12. Weiser, M., Gold, R., Brown, J.S.: The origins of ubiquitous computing research at parc in the late 1980s. IBM Syst. J. 38, 693–696 (1999)
13. Bell, G., Dourish, P.: Yesterday's tomorrows: notes on ubiquitous computing's dominant vision. Personal Ubiquitous Comput. 11, 133–143 (2007)
14. Weiser, M.: Ubiquitous computing. Computer 26, 71–72 (1993)
15. Weiser, M.: Some computer science issues in ubiquitous computing. Commun. ACM 36, 75–84 (1993)
16. Weiser, M.: Keynote: Building invisible interfaces (1994)
17. Abowd, G.D., Mynatt, E.D., Rodden, T.: The human experience. IEEE Pervasive Computing 1, 48–57 (2002)
18. Abowd, G.D., Mynatt, E.D.: Charting past, present, and future research in ubiquitous computing. ACM Trans. Comput.-Hum. Interact. 7, 29–58 (2000)
19. Silva, T.H., de S. Celes, C.S.F., Mota, V.F.S., Loureiro, A.A.F.: A picture of present ubicomp research exploring publications from important events in the field. Journal of Applied Computing Research 2(1), 32–49 (2012)
20. Schilit, B., Adams, N., Want, R.: Context-aware computing applications. In: Proceedings of the 1994 First Workshop on Mobile Computing Systems and Applications, pp. 85–90. IEEE Computer Society, Washington, DC (1994)
21. Dey, A.K., Abowd, G.D., Wood, A.: Cyberdesk: a framework for providing self-integrating context-aware services. In: Proceedings of the 3rd International Conference on Intelligent User Interfaces, IUI 1998, pp. 47–54. ACM, New York (1998)
22. Pascoe, M.J.: Adding generic contextual capabilities to wearable computers. In: Proceedings of the 2nd IEEE International Symposium on Wearable Computers, ISWC 1998, pp. 92–99. IEEE Computer Society, Washington, DC (1998)
23. Dey, A.K., Abowd, G.D.: Towards a Better Understanding of Context and Context-Awareness. In: CHI 2000 Workshop on the What, Who, Where, When, and How of Context-Awareness (2000)
24. Mashhadi, A.J., Capra, L.: Quality Control for Real-time Ubiquitous Crowdsourcing. In: Proc. of the 2nd Int'l Workshop on Ubiquitous Crowdsouring (UbiCrowd 2011), pp. 5–8 (2011)

25. Silva, T.H., Vaz de Melo, P.O.S., Almeida, J.M., Salles, J., Loureiro, A.A.F.: A picture of Instagram is worth more than a thousand words: Workload characterization and application. In: Proc. of the IEEE International Conference on Distributed Computing in Sensor Systems (DCOSS 2013) (May 2013)
26. Long, X., Jin, L., Joshi, J.: Exploring trajectory-driven local geographic topics in foursquare. In: Proceedings of the 2012 ACM Conference on Ubiquitous Computing. UbiComp 2012, pp. 927–934. ACM, New York (2012)
27. Giannotti, F., Pedreschi, D., Pentland, A., Lukowicz, P., Kossmann, D., Crowley, J., Helbing, D.: A planetary nervous system for social mining and collective awareness. The European Physical Journal Special Topics 214(1), 49–75 (2012)
28. Brockmann, D., Hufnagel, L., Geisel, T.: The scaling laws of human travel. Nature 439(7075), 462–465 (2006)
29. Zheng, Y., Zhang, L., Xie, X., Ma, W.Y.: Mining interesting locations and travel sequences from gps trajectories. In: Proceedings of the 18th International Conference on World Wide Web, pp. 791–800. ACM (2009)
30. Gonzalez, M.C., Hidalgo, C.A., Barabasi, A.L.: Understanding individual human mobility patterns. Nature 453(7196), 779–782 (2008)
31. Song, C., Qu, Z., Blumm, N., Barabási, A.L.: Limits of predictability in human mobility. Science 327(5968), 1018–1021 (2010)
32. Cheng, Z., Caverlee, J., Lee, K., Sui, D.Z.: Exploring Millions of Footprints in Location Sharing Services. In: Proc. of the Fifth Int'l Conf. on Weblogs and Social Media, ICWSM 2011 (2011)
33. Cho, E., Myers, S.A., Leskovec, J.: Friendship and mobility: user movement in location-based social networks. In: Proceedings of the 17th ACM SIGKDD International Conference on Knowledge Discovery and Data Mining, KDD 2011, pp. 1082–1090. ACM, San Diego (2011)
34. Nguyen, T., Szymanski, B.K.: Using location-based social networks to validate human mobility and relationships models. arXiv preprint arXiv:1208.3653 (2012)
35. Zheng, Y.T., Zha, Z.J., Chua, T.S.: Mining travel patterns from geotagged photos. ACM Trans. Intell. Syst. Technol. 3(3), 56:1–56:18 (2012)
36. Bilandzic, M., Foth, M.: A review of locative media, mobile and embodied spatial interaction. International Journal of Human-Computer Studies 70(1), 66–71 (2012)
37. Cranshaw, J., Schwartz, R., Hong, J.I., Sadeh, N.: The Livehoods Project: Utilizing Social Media to Understand the Dynamics of a City. In: Proc. of the Sixth Int'l Conf. on Weblogs and Social Media (2012)
38. Ng, A.Y., Jordan, M.I., Weiss, Y., et al.: On spectral clustering: Analysis and an algorithm. Advances in Neural Information Processing Systems 2, 849–856 (2002)
39. Noulas, A., Scellato, S., Mascolo, C., Pontil, M.: Exploiting Semantic Annotations for Clustering Geographic Areas and Users in Location-based Social Networks. In: Proc. of the Fifth Int'l Conf. on Weblogs and Social Media, ICWSM 2011 (2011)
40. Luxburg, U.: A tutorial on spectral clustering. Statistics and Computing 17(4), 395–416 (2007)
41. Blei, D.M., Ng, A.Y., Jordan, M.I.: Latent dirichlet allocation. The Journal of Machine Learning Research 3, 993–1022 (2003)
42. Silva, T.H., Vaz de Melo, P.O.S., Almeida, J.M., Salles, J., Loureiro, A.A.F.: Visualizing the invisible image of cities. In: Proc. of IEEE International Conference on Cyber, Physical and Social Computing, CPScom 2012 (November 2012)
43. Kisilevich, S., Krstajic, M., Keim, D., Andrienko, N., Andrienko, G.: Event-based analysis of people's activities and behavior using flickr and panoramio geotagged photo collections. In: IEEE 2010 14th International Conference on Information Visualisation (IV), pp. 289–296 (2010)

44. Ester, M., Kriegel, H.P., Sander, J., Xu, X.: A density-based algorithm for discovering clusters in large spatial databases with noise, Kdd (1996)
45. Frias-Martinez, V., Soto, V., Hohwald, H., Frias-Martinez, E.: Characterizing urban landscapes using geolocated tweets. In: IEEE 2012 International Conference on and 2012 International Confernece on Social Computing (SocialCom) Privacy, Security, Risk and Trust (PASSAT), pp. 239–248 (2012)
46. Kohonen, T.: The self-organizing map. Proceedings of the IEEE 78(9), 1464–1480 (1990)
47. Cheng, Y.: Mean shift, mode seeking, and clustering. IEEE Transactions on Pattern Analysis and Machine Intelligence 17(8), 790–799 (1995)
48. Ji, R., Xie, X., Yao, H., Ma, W.Y.: Mining city landmarks from blogs by graph modeling. In: Proceedings of the 17th ACM International Conference on Multimedia, pp. 105–114. ACM (2009)
49. Wang, C., Xie, X., Wang, L., Lu, Y., Ma, W.Y.: Detecting geographic locations from web resources. In: Proceedings of the 2005 Workshop on Geographic Information Retrieval, pp. 17–24. ACM (2005)
50. Nister, D., Stewenius, H.: Scalable recognition with a vocabulary tree. In: 2006 IEEE Computer Society Conference on Computer Vision and Pattern Recognition, vol. 2, pp. 2161–2168. IEEE (2006)
51. Kleinberg, J.M.: Authoritative sources in a hyperlinked environment. Journal of the ACM (JACM) 46(5), 604–632 (1999)
52. Scellato, S., Noulas, A., Lambiotte, R., Mascolo, C.: Socio-spatial Properties of Online Location-based Social Networks. In: Proc. of the Fifth Int'l Conf. on Weblogs and Social Media, ICWSM 2011 (2011)
53. Cranshaw, J., Toch, E., Hong, J., Kittur, A., Sadeh, N.: Bridging the gap between physical location and online social networks. In: Proceedings of the 12th ACM International Conference on Ubiquitous Computing, pp. 119–128. ACM (2010)
54. Quercia, D., Capra, L., Crowcroft, J.: The social world of twitter: Topics, geography, and emotions. In: The 6th International AAAI Conference on Weblogs and Social Media, Dublin (2012)
55. Burt, R.S.: Structural holes: The social structure of competition (1992)
56. Java, A., Joshi, A., Finin, T.: Detecting commmunities via simultaneous clustering of graphs and folksonomies. In: Proceedings of WebKDD, vol. 2008 (2008)
57. Shi, J., Malik, J.: Normalized cuts and image segmentation. IEEE Transactions on Pattern Analysis and Machine Intelligence 22(8), 888–905 (2000)
58. Sadilek, A., Kautz, H., Bigham, J.P.: Finding your friends and following them to where you are. In: Proceedings of the Fifth ACM International Conference on Web Search and Data Mining, pp. 723–732. ACM (2012)
59. Bollen, J., Mao, H., Zeng, X.: Twitter mood predicts the stock market. Journal of Computational Science 2(1), 1–8 (2011)
60. Sakaki, T., Okazaki, M., Matsuo, Y.: Earthquake shakes twitter users: real-time event detection by social sensors. In: Proceedings of the 19th International Conference on World Wide Web, WWW 2010, pp. 851–860. ACM, New York (2010)
61. Lee, R., Sumiya, K.: Measuring geographical regularities of crowd behaviors for twitter-based geo-social event detection. In: Proceedings of the 2nd ACM SIGSPATIAL International Workshop on Location Based Social Networks, pp. 1–10. ACM (2010)
62. McGill, R., Tukey, J.W., Larsen, W.A.: Variations of box plots. The American Statistician 32(1), 12–16 (1978)
63. Becker, H., Naaman, M., Gravano, L.: Beyond trending topics: Real-world event identification on twitter. In: Proceedings of the Fifth International AAAI Conference on Weblogs and Social Media, ICWSM 2011 (2011)

64. Joseph, K., Tan, C.H., Carley, K.M.: Beyond local, categories and friends: clustering foursquare users with latent topics. In: Proceedings of the 2012 ACM Conference on Ubiquitous Computing, pp. 919–926. ACM (2012)
65. Naaman, M., Zhang, A.X., Brody, S., Lotan, G.: On the study of diurnal urban routines on twitter. In: Sixth International AAAI Conference on Weblogs and Social Media (2012)
66. Poblete, B., Garcia, R., Mendoza, M., Jaimes, A.: Do all birds tweet the same?: characterizing twitter around the world. In: Proceedings of the 20th ACM International Conference on Information and Knowledge Management, pp. 1025–1030. ACM (2011)
67. Bradley, M.M., Lang, P.J.: Affective norms for english words (anew): Instruction manual and affective ratings. Technical report, Technical Report C-1, The Center for Research in Psychophysiology, University of Florida (1999)
68. Redondo, J., Fraga, I., Padrón, I., Comesaña, M.: The spanish adaptation of anew (affective norms for english words). Behavior Research Methods 39(3), 600–605 (2007)
69. Gao, H., Tang, J., Liu, H.: gscorr: modeling geo-social correlations for new check-ins on location-based social networks. In: Proceedings of the 21st ACM International Conference on Information and Knowledge Management, pp. 1582–1586. ACM (2012)
70. Yu, H., Sun, G., Lv, M.: Users sleeping time analysis based on micro-blogging data. In: Proceedings of the 2012 ACM Conference on Ubiquitous Computing, pp. 964–968. ACM (2012)
71. Saroiu, S., Wolman, A.: I am a sensor, and i approve this message. In: Proc. of the Eleventh Workshop on Mobile Computing Systems and Applications, HotMobile 2010, pp. 37–42. ACM, Annapolis (2010)
72. Reddy, S., Estrin, D., Hansen, M., Srivastava, M.: Examining micro-payments for participatory sensing data collections. In: Proc. of the 12th ACM International Conference on Ubiquitous Computing, Ubicomp 2010, pp. 33–36. ACM, New York (2010)
73. Pontes, T., Magno, G., Vasconcelos, M., Gupta, A., Almeida, J., Kumaraguru, P., Almeida, V.: Beware of what you share: Inferring home location in social networks. In: 2012 IEEE 12th International Conference on Data Mining Workshops (ICDMW), pp. 571–578 (2012)
74. Toch, E., Cranshaw, J., Drielsma, P.H., Tsai, J.Y., Kelley, P.G., Springfield, J., Cranor, L., Hong, J., Sadeh, N.: Empirical models of privacy in location sharing. In: Proceedings of the 12th ACM International Conference on Ubiquitous Computing, Ubicomp 2010, pp. 129–138. ACM, Copenhagen (2010)
75. Brush, A.B., Krumm, J., Scott, J.: Exploring end user preferences for location obfuscation, location-based services, and the value of location. In: Proceedings of the 12th ACM International Conference on Ubiquitous Computing, Ubicomp 2010, pp. 95–104. ACM, Copenhagen (2010)
76. Kim, D.H., Han, K., Estrin, D.: Employing user feedback for semantic location services. In: Proc. of the 13th International Conference on Ubiquitous Computing, UbiComp 2011, pp. 217–226. ACM, New York (2011)
77. Hsieh, H.P., Li, C.T., Lin, S.D.: Exploiting large-scale check-in data to recommend time-sensitive routes. In: Proc. of the ACM SIGKDD Int Workshop on Urban Computing, UrbComp 2012, pp. 55–62. ACM, Beijing (2012)
78. Silva, T.H., Vaz de Melo, P.O.S., Almeida, J.M., Loureiro, A.A.F.: Challenges and opportunities on the large scale study of city dynamics using participatory sensing. In: 18th IEEE Symposium on Computers and Communications (ISCC 2013), Split, Croatia (July 2013)

An Analysis of Interactions within and between Extreme Right Communities in Social Media

Derek O'Callaghan[1], Derek Greene[1], Maura Conway[2], Joe Carthy[1],
and Pádraig Cunningham[1]

[1] School of Computer Science & Informatics, University College Dublin
{derek.ocallaghan,derek.greene,joe.carthy,padraig.cunningham}@ucd.ie
[2] School of Law & Government, Dublin City University
maura.conway@dcu.ie

Abstract. Many extreme right groups have had an online presence for some time through the use of dedicated websites. This has been accompanied by increased activity in social media platforms in recent years, enabling the dissemination of extreme right content to a wider audience. In this paper, we present an analysis of the activity of a selection of such groups on Twitter, using network representations based on reciprocal follower and interaction relationships, while also analyzing topics found in their corresponding tweets. International relationships between certain extreme right groups across geopolitical boundaries are initially identified. Furthermore, we also discover stable communities of accounts within local interaction networks, in addition to associated topics, where the underlying extreme right ideology of these communities is often identifiable.

Keywords: network analysis, social media, community detection, topic modelling, Twitter, extreme right.

1 Introduction

Groups associated with the extreme right have maintained an online presence for some time [1,2], where dedicated websites have been employed for the purposes of content dissemination and member recruitment. Recent years have seen increased activity by these groups in social media platforms, given the potential to access a far wider audience than was previously possible [3,4]. In this paper, we present an analysis of the activity of a selection of these groups on Twitter, where the focus is upon groups of a fascist, racist, supremacist, extreme nationalist or neo-Nazi nature, or some combination of these. Twitter's features enable extreme right groups to disseminate hate content with relative ease, while also facilitating the formation of communities of accounts around variants of extreme right ideology. Message posts (*tweets*) by members of these groups, to which access is usually unrestricted, are often used to redirect users to content hosted on external websites, including dedicated websites managed by particular groups, or content sharing platforms such as YouTube.

M. Atzmueller et al. (Eds.): MUSE/MSM 2012, LNAI 8329, pp. 88–107, 2013.
© Springer-Verlag Berlin Heidelberg 2013

For the purpose of this analysis, we have retrieved data for a selection of identified extreme right Twitter accounts from eight countries. Our initial objective is the identification of international relationships between certain groups that transcend geopolitical boundaries. This involves the analysis of two network representations of the accounts from the eight country sets, based on reciprocal follower and interaction relationships. Here, interactions are derived from observations of mentions and retweets between accounts. It appears that a certain amount of international awareness exists between accounts based on the follower relationship, while interactions indicate stronger relationships where linguistic and geographical proximity are highly influential.

This leads to our next objective of analyzing communities of extreme right accounts found within local interaction networks, where locality is considered in terms of nationality or linguistic proximity. Tweet content is also analyzed for the purpose of generating interpretable descriptions for the detected communities, in addition to the discovery of latent topics associated with interactions between the member accounts. We find that matrix factorization techniques are more suitable for topic analysis of these particular data sets. By using the same account profile document representation for both community description generation and topic discovery, it is possible to generate a mapping between the detected communities and their associated topics. Each community description and corresponding topic mapping can then be used in conjunction with manual analysis of the account profiles, tweets and external websites to provide an interpretation of the underlying community ideology. While we observe some community division along electoral and non-electoral lines, this is not clear in all cases. Other notable findings include communities of a more traditional conservative nature, opposition to bodies such as the EU, and the influence of concerns such as counter-Jihad on international relationships.

In Section 2, we provide a description of related work based on the online activities of extremist groups. The collection of the Twitter data sets is then discussed in Section 3. Analysis of the international relationships between extreme right groups from the eight countries is presented in Section 4. Next, in Section 5, we describe the discovery of extreme right communities within local interaction networks, including the methodology used for network derivation, community detection, stability ranking, description generation and topic analysis. We focus on two case studies using English and German language networks, where we offer an interpretation of a selection of these communities. Finally, the overall conclusions are discussed in Section 6, and some suggestions for future work are made.

2 Related Work

The online activities of different varieties of extremist groups including those associated with the extreme right have been the subject of a number of studies. Burris et al. used social network analysis to study a network based on the links between a selection of white supremacist websites [2]. They found this network to be relatively decentralized with multiple centres of influence, while also appearing to be mostly undivided along doctrinal lines. Similar decentralization and multiple communities were found by Chau and Xu in their study of networks built from users contributing to hate group and racist blogs [5]. They also found that some of these groups exhibited transnational characteristics. In a similar approach to that of Burris et al., Tateo analyzed groups associated with the Italian extreme right, using networks based on links between group websites [6]. Caiani and Wagemann studied similar Italian groups along with those from the German extreme right, where they found the German network to be structurally centralized to a greater extent than that of the Italian groups [7]. The contents of websites belonging to central nodes within Russian extreme right networks were analyzed by Zuev [8]. In their review of the conservative movement in the USA, Blee and Creasap [9] discuss the engagement in online activity as part of an overall mobilization strategy by the more extremist groups within it.

The potential for online radicalization through exposure to jihadi video content on YouTube was investigated by Bermingham et al., where it was suggested that a potentially increased online leadership role may be attributed to users claiming to be women, according to centrality, network density and average speed of communication [10]. Sureka et al. also studied the activity of extremist users within YouTube, investigating content properties along with hidden network communities [11]. Bartlett et al. performed a survey of European populist party and group supporters on Facebook [3], while Baldauf et al. also investigated the use of Facebook by the German extreme right [4]. As the majority of this work involved the study of dedicated websites managed by extreme right groups, we believed that an analysis of their activity in social media, focusing specifically on extreme right Twitter communities, would complement this by providing additional insight into the overall online presence of these groups.

3   Data

Twitter data was collected to facilitate the analysis of contemporary online extreme right activity. We identified initial sets of relevant accounts for a selection of countries, where the country selection was informed by prior knowledge of extreme right groups. Based on earlier studies [4,7], the criteria used to identify relevant accounts included profiles containing references to known groups or employing extreme right symbols; recent tweet activity; similar Facebook/YouTube accounts; reciprocal follower relationships with known relevant accounts; accounts with self-curated Twitter lists containing relevant accounts; extreme right media accounts such as record labels and concert organisers. Details of these country sets can be found in Table 1.

Certain accounts were not included, such as inactive accounts, or those that were not deemed to be related to the extreme right. These included traditional conservative (e.g. centre-right) accounts, non-conformists/anti-establishment accounts considered to be left-wing, and conspiracy theorists. As we were initially interested in non-electoral extreme right groups [12], higher-profile politicians or

Table 1. Data set sizes for eight countries of interest

Country	Number of Accounts
France	25
Germany	53
Greece	45
Italy	17
Spain	43
Sweden	21
UK	32
USA	32

political parties were ignored for the most part, with a minor number of these accounts included where it was felt that there was a close association with relevant accounts. An obstacle was the language barrier, where the use of online translation tools did not always prove helpful in the interpretation of ambiguous profiles. In cases where the relevance of an account profile was inconclusive, that account was ignored. Twitter data including followers, friends, tweets and list memberships were retrieved for each of the selected accounts during the period March – August 2012, as limited by the Twitter API restrictions effective at the time.

4   International Relationships

We began with an analysis of the international relationships between the identified extreme right groups, based on interactions between the accounts from the eight country sets. An *interaction* is defined as one account "mentioning" another account within a tweet, or an account "retweeting" a tweet generated by another account. Both types of event were included in order to address issues of data sparsity and incompleteness. We were particularly interested in reciprocal activity between accounts, where such activity can potentially indicate the presence of a stronger relationship. For example, previous work has used reciprocal mentions between accounts to represent dialogue [13]. An *interactions network* was created with n nodes representing accounts, and m undirected weighted edges representing reciprocal mentions and retweets between pairs of accounts, with weights corresponding to the number of interactions. All observed interactions found in the retrieved data sets were considered. Any connected components of size < 5 were filtered. In addition, we were also interested in international follower relationships, and a similar undirected unweighted network was created to capture reciprocal follower links between the accounts in different country sets. Throughout this analysis, due to the sensitivity of the subject matter, and in the interest of privacy, individual accounts are not identified; instead, we restrict discussion to known extreme right groups and their affiliates.

4.1   International Follower Awareness

The international followers network can be seen in Fig. 1. As might be expected, most of the follower relationships are between accounts from the same country, although a certain number of international relationships are evident. It would appear that linguistic and geographical proximity is influential here, for example, we can observe relationships between the Spanish and Italian (and to a lesser extent, French) accounts, with strong connections also between the UK and USA. Similar behaviour with respect to social ties in Twitter has been identified by Takhteyev et al. and Kulshrestha et al. [14,15]. However, there appear to be some exceptions to the influence of geographical proximity, most notably, Swedish (yellow) and Italian (green) accounts that are not co-located with their respective country nodes. In both cases, the majority of tweets from these accounts are in English, which presumably ensures a wider audience. The former account is a Swedish representative of a pan-Scandinavian group espousing national socialist ideas, who appears to be interacting with many international accounts, particularly from the USA. The Italian account is a national socialist whose tweets often contain URLs to music or video content hosted on external websites, but it is unclear if a direct association exists with any particular group. From an analysis of other central nodes in the network (using betweenness centrality), it would seem that those involved in the dissemination of material via external URLs, or media platforms such as extreme right news websites and

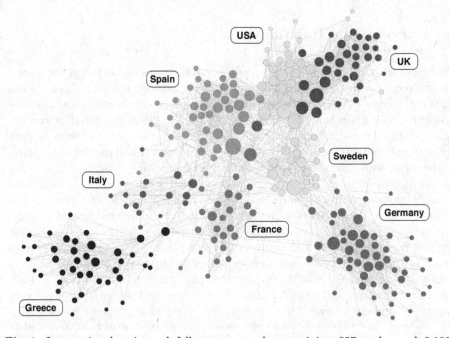

Fig. 1. International reciprocal followers network, containing 257 nodes and 2,100 edges. Node size is proportional to degree.

radio stations, are attempting to raise awareness amongst a variety of international followers. This is particularly the case when the English language is used.

4.2 International Interactions

The follower-based relationship between international accounts could be considered as passive when compared with that of the interaction networks, where such interactions can be indicative of actual dialogue between accounts. The interactions network in Fig. 2 is somewhat smaller than the corresponding network in Fig. 1. We also observe that the Greek community is now disconnected from the largest connected component. Apart from this, the network has a similar structure to that of the followers network, in that most interaction occurs within individual country-based communities. Connections between these communities do exist, but are fewer than in the followers network. The influence of linguistic proximity appears to take precedence here, with the use of English playing a major role as mentioned in the previous section. For example, a relatively large number of connections remain present between accounts in the UK and USA. In the case of the German community, while the followers network contains a variety of connections with other international accounts, this has now been reduced to connections between two German accounts and a small number of UK accounts, in addition to an account acting as an English language Twitter channel for a Swedish nationalist group. Similarly, the Swedish account co-located with the USA community is the same account as that in the followers network, who appears to be involved in many English-based interactions with international accounts.

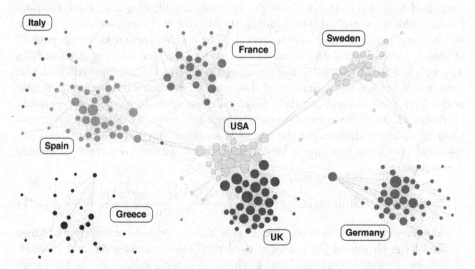

Fig. 2. International reciprocal interactions network, containing 218 nodes and 1,186 edges. Node size is proportional to degree.

5   Local Analysis

Following the analysis of international relationships described in the previous section, we then proceeded to analyze interactions within local networks with the objective of detecting specific communities of related accounts, focusing on two data sets as case studies. Based on our observation of linguistic proximity at the international level, we merged the UK and USA data sets to produce a single English language data set for one case study, while the second focused on the German language data set. In both cases, we created *expanded* versions of the initial data sets to facilitate a more detailed analysis, where all available data were also retrieved for those accounts having a reciprocal follower relationship with more than one of the original identified accounts. As Twitter follower relationships tend to exhibit lower reciprocity than other social networking sites [16], the understanding was that this action would be largely isolated to accounts having a relatively stronger relationship with those from the original data sets. This process resulted in the inclusion of 1,513 and 448 accounts respectively in the expanded English and German language data sets.

5.1   Community Detection in Interaction Networks

For the purpose of community detection based on account interactions, we constructed undirected interaction networks (G) from the expanded data sets. As with the international analysis, only reciprocal edges were used in order to capture stronger relationships, all observed interactions were included, and connected components of size < 5 were filtered. We used our variant of the work by Lancichinetti & Fortunato to generate a set of stable *consensus communities* from such a network [17,18], where 100 runs of the OSLOM algorithm [19] were used to generate the consensus communities. Following this, the consensus communities were ranked based on the stability of their members with respect to the corresponding consensus matrix M. For a given consensus community C of size c, we computed the mean of the values M_{xy} for all unique pairs (L_x, L_y) assigned to C; this value has the range $[0, 1]$. We then computed the mean expected value for a community of size c as follows: randomly select c unique nodes from G, and compute their mean pairwise score from the corresponding entries in M. This process was repeated over a large number of randomised runs, yielding an approximation of the expected stability value. The widely-used adjustment technique introduced by [20] was then employed to correct for chance agreement:

$$\text{CorrectedStability}(C) = \frac{\text{Stability}(C) - \text{ExpectedStability}(C)}{1 - \text{ExpectedStability}(C)} \quad (1)$$

A value close to 1 will indicate that C is a highly-stable community. As higher values of the threshold parameter τ used with the consensus method resulted in sparser consensus networks, and having tested with values of τ in the range $[0.1, 0.9]$, we selected $\tau = 0.5$ as a compromise between node retention and more stable communities. The resolution of communities found by OSLOM is directly controlled by the associated parameter P, which has a default value of 0.5 in the implementation. Although Lancichinetti et al. state that $P = 0.1$ delivered an excellent performance on the benchmark graphs used in the OSLOM paper, they also suggest that it would be more appropriate to estimate P on a case by case basis [19]. We found that using $P = 0.1$ tended to detect a larger number of relatively small communities. Increasing values of P ($[0.1, 0.9]$) produced smaller numbers of larger communities having higher stability scores (with $\tau = 0.5$). Given this, for both data sets, we used the corresponding value of P that produced the highest mean stability score to generate communities for detailed analysis. Further details of the P values used can be found later in the case study sections. For the purpose of this analysis, we focus on larger consensus communities having ≥ 10 members.

5.2   Community Descriptions

The content of tweets was also analyzed for the purpose of (1) generating interpretable descriptions for the detected communities, and (2) identifying latent topics associated with interactions between the accounts assigned to these communities. Following the approach of Hannon et al. [21], we generated a "profile document" for each node in an interactions network, consisting of an aggregation of their corresponding tweets, from which a tokenized representation was produced. Our initial experiments used all available terms in the account documents, where the tokenization process involved the exclusion of URLs and stopwords (additional social networking stopwords such as "ff", "facebook" etc. were included with multiple language stopword lists), normalization of diacritics, and stemming of the remaining terms. Low-frequency terms (appearing in < 4 account documents) and documents containing < 10 terms were excluded. These profile documents were then represented by log-based TF-IDF term vectors, which were subsequently normalized to unit length.

However, we encountered two issues with the use of all available tweet terms as candidates for this representation. It transpired that the expansion of this data set through the addition of reciprocal follower accounts of the original accounts resulted in the inclusion of accounts whose mother tongue was not English, for example, accounts from South Africa and Sweden. As a result, generic non-English terms were treated as highly discriminating due to their relative low frequency within the full set of English terms across all documents. Separately, the use of all terms required extensive maintenance of a multilingual stopword list. To address these issues, we excluded all terms apart from hashtags when generating the account documents, which greatly reduced the number of required stopwords, while also promoting more discriminating non-English terms in subsequent analysis. As it was possible to generate hashtag-based account documents for 96% of English language accounts and 92% of German language accounts present in the corresponding interaction networks, it was felt that sufficient coverage of the accounts was retained. In addition, the dimensionality of the vector representations was also considerably reduced by the sole use of hashtags. Having produced these TF-IDF vectors, each community description was generated by selecting the subset of vectors for the accounts assigned to the community, and calculating a mean vector D from this subset matrix. The final community description consisted of the top ten hashtags from D. For the remainder of this paper, all hashtags are presented without the preceding "#" character.

5.3   Topic Analysis

Topic modelling is concerned with the discovery of latent semantic structure or topics within a set of documents, which can be derived from co-occurrences of words and documents [22]. This strategy dates back to the early work on latent semantic indexing by Deerwester et al. [23]. Popular methods include probabilistic models such as latent Dirichlet allocation (LDA) [24], or matrix factorization techniques such as Non-negative matrix factorization (NMF) [25]. These have previously been successfully applied in social media analytics, for example, the work of Ramage et al. [26], Weng et al. [27], and Saha and Sindhwani [28]. We initially evaluated both LDA and NMF-based methods with the account document representations described above, where NMF was found to produce the most readily-interpretable results. This appeared to be due to the tendency of LDA to discover topics that over-generalized [29].

The observed connections between the countries at the international level suggested that specific topics associated with smaller groups of accounts were present in the extended English and German language data sets. Given this, in the trade-off between generality and specificity, we opted for the latter in this particular analysis, and so used NMF with the same TF-IDF account vectors as before for topic analysis. Here, the IDF component ensured a lower ranking for less discriminating terms, thus leading to the discovery of more specific topics. Having constructed an $n \times m$ term-document matrix V, where each column contained a TF-IDF account vector, NMF produced two factors; W, an $n \times T$ matrix containing topic basis vectors and H, a $T \times m$ matrix containing the topic assignments for each account document. To address the instability introduced by random initialization in standard NMF, we employed the NNDSVD method proposed by Boutsidis and Gallopoulos [30], which is particularly suitable for sparse matrices.

Although the interactions network topology and general topic analysis provided two different views on a particular data set, in this analysis, we were primarily interested in the communities within the interactions networks, where the associated topics could be used for further interpretation of the account membership. As the same hashtag-based account document TF-IDF vectors were used for both community description generation and topic detection, it was possible to generate a mapping between the detected communities and topics. For each community, we ranked the topics according to the cosine similarity between its mean community hashtag description vector D and the topic basis vectors W. This method occasionally detected multiple similar topics for a particular community, which was to be expected given that individual account documents could themselves be associated with multiple topics. To avoid redundant mappings, we ignored any topics having a cosine similarity < 0.1 with each D vector, as using this threshold appeared to produce relevant community-topic mappings in general for both data sets.

5.4 Case Study: English language

An interactions network (Fig. 3) was created for the English language data set, consisting of 1,034 nodes and 9,429 edges. Due to the effect of the resolution parameter P on the communities found by OSLOM, we measured the mean stability score ($\tau = 0.5$) and number of communities found for values of P in $[0.1, 0.9]$, using the consensus community method previously described. A plot of the mean score and sizes can be found in Fig. 4, where it can be seen that the highest mean score was generated for $P = 0.4$, while the number of communities found had also stabilized at ~ 34. Eight communities having at least ten members were found for $P = 0.4$ (86% of total network account nodes), and their corresponding hashtag descriptions and stability scores can be found in Table 2. These communities have been manually annotated with identifiers for reference in the subsequent discussion, based on an analysis of the account member composition (for example, EAST, LONDON). Topic analysis was also performed using NMF, with number of topics $T = 15$. Having experimented with a range of values for T, we used $T = 15$ as this was the smallest value which led to the emergence of non-English language topics.

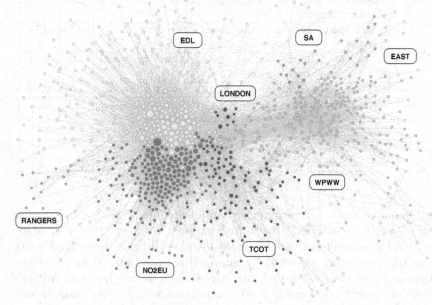

Fig. 3. English language interactions network with 1,034 nodes and 9,429 edges. Node size is proportional to degree. Consensus communities ($P = 0.4$, #members ≥ 10) are labelled.

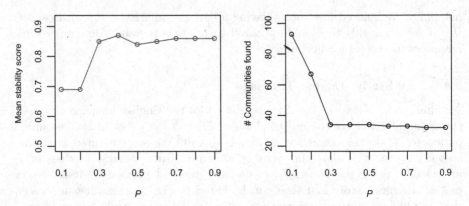

Fig. 4. English language mean stability scores (left) and number of communities found (right) for P in $[0.1, 0.9]$

Table 2. Consensus communities found in the English language interactions network ($\tau = 0.5$, $P = 0.4$, #members ≥ 10), representing 86% of all network nodes

Id	Description	Size	Score
SA	anc, onswiloorleef, southafrica, ancyl, onssaloorleef, zuma, stopabsa, zumaspear, nuus, svpol	13	1.00
EAST	presstv, sv, wpww, metal, estonia, 666, polska, ww2, edl, ukraine	11	0.95
WPWW	wpww, tcot, whitepower, nigger, p2, niggers, teaparty, gop, obama, whitepride	240	0.88
TCOT	tcot, teaparty, p2, obama, gop, israel, tlot, lnyhbt, sgp, islam	91	0.74
NO2EU	no2eu, ukip, bbcqt, labour, edl, leveson, newsnight, euro2012, eurovision, london2012	214	0.74
EDL	edl, uaf, islam, bbcqt, lfc, bnp, muslim, tcot, mufc, israel	290	0.62
RANGERS	watp, rangersfamily, nosurrender, gersfollowback, wedontdowalkingaway, rfc, taintedtitle, rangersfamilly, rangers, rangerfamily	16	0.58
LONDON	edl, coys, stgeorgesday, londonriots, whys, savages, bbcqt, stfc, tottenham, dench	10	0.53

The accounts in the SA community appear to be white South Africans, with some profiles containing racist and national socialist references. Many tweets from these accounts relate to perceived cultural threats from black South Africans that are often retweeted by international accounts. Most of the description hashtags are related to South Africa, such as *anc* (*African National Congress*, the current governing party) and *onswiloorleef* ("We want to survive"), associated with a campaign highlighting alleged violent attacks against Afrikaners. This community is associated with a single South African topic, which is to be expected given the discriminating nature of these hashtags.

The EAST community includes white power/national socialist accounts from Eastern European countries such as Estonia, Poland and Ukraine, who occasionally tweet in English. Notable description hashtags include *wpww* (white pride world wide) and *metal*, where the latter refers to a music sub-genre known as nationalist socialist black metal. Of similar interest is *presstv*, the Iranian state-owned English language news network. This has previously been accused of propagating anti-Semitic content, while also hosting Holocaust-deniers and white nationalists. The most similar topic for this community is one connected to white pride ideology that includes media references, for example, *wpradio* (white pride radio).

The WPWW community would appear to be national socialist/white power in nature, with the appearance of hashtags such as *wpww* and *whitepower*. An analysis of the accounts and associated profiles finds references to the *American*

Table 3. English language communities and associated NMF topics ($T = 15$, hashtag description cosine similarity ≥ 0.1)

Community	Similarity	Top 10 Topic Terms
SA	0.56	anc, onswiloorleef, stopabsa, ancyl, southafrica, zuma, nuus, afrikaans, zumaspear, genocide
EAST	0.12	wpww, whitepride, wpradio, contest, staywhite, whiteisright, white, whiteunity, genocide, skinhead
WPWW	0.14	wpww, whitepride, wpradio, contest, staywhite, whiteisright, white, whiteunity, genocide, skinhead
	0.10	tcot, teaparty, p2, gop, tlot, obama, sgp, ocra, lnyhbt, twisters
TCOT	0.48	tcot, teaparty, p2, gop, tlot, obama, sgp, ocra, lnyhbt, twisters
	0.17	israel, islam, sharia, muslim, jihad, iran, gaza, syria, egypt, nadarkhani
	0.15	prolife, abortion, tcot, prochoice, personhood, god, octoberbaby, 912, gingrich, preborn
NO2EU	0.16	labour, leveson, occupylsx, cameron, ukuncut, bnp, greece, bbc, syria, olympics
	0.13	bbcqt, newsnight, eurovision, london2012, euro2012, closingceremony, pmqs, teamgb, leveson, olympics
	0.12	no2eu, lab11, english, labour, england, eurozone, euro, eurocrash, euro2012, london2012
	0.12	ukip, voteukip, christappin, uk, greece, ronpaul, freegary, tories, richardo, extradition
EDL	0.42	edl, uaf, islam, bnp, casualsunited, luton, rochdale, mdl, bristol, praetorian
RANGERS	0.59	rangersfamily, watp, nosurrender, rangers, gersfollowback, rfc, wedontdowalkingaway, taintedtitle, rangersfamilly, celtictaintedtitle
LONDON	0.15	edl, uaf, islam, bnp, casualsunited, luton, rochdale, mdl, bristol, praetorian

Nazi Party, along with other related terms such as *14* (a reference to a 14-word slogan coined by the white supremacist David Lane), and *88* ("heil hitler") in account names. There are also references to skinhead groups, including a website where related media and merchandise can be found for sale. Accounts appear to be mostly associated with the USA, although a small number of European accounts are also present. References to *tcot* (top conservatives on Twitter) and *gop* (US Republican Party) can also be seen, indicating the presence of more traditional conservative accounts. However, this does not necessarily point to any official link between these groups. Two topics are most closely associated with this community, mirroring the account and description hashtag division between *wpww* and *tcot*. The TCOT community is largely composed of traditional conservatives, where the description contains a number of hashtags commonly used by these groups such as *gop* and *teaparty*. However, we also note the presence of *p2* (progressives, Tweet Progress), effectively the polar opposite of *tcot*. A number of anti-Islamic counter-Jihad accounts are also present [31]. This community is strongly linked to the *tcot* topic, with the counter-Jihad and pro-life/anti-abortion topics also of interest.

Opposition to the EU appears to be the binding theme of the NO2EU community, which contains several political or electoral accounts, including a number affiliated with British Eurosceptic parties such as the *United Kingdom Independence Party* (UKIP). Non-electoral British nationalist accounts are also present, where their tweets and profiles often contain anti-EU statements and imagery. We also see references to British media such as BBC current affairs programmes (*bbcqt*, *newsnight*). Accordingly, topics linked to this community appear to be concerned with politics and the EU in general. The EDL community consists mostly of accounts associated with the *English Defence League* (*edl*), a counter-Jihad movement opposed to the alleged spread of radical Islamism within the UK [31,32]. Other accounts include those associated with *Casuals United*, a protest group linked with the EDL that formed from an alliance of football hooligans (references to football clubs can also be observed). The *uaf* hashtag refers to the *Unite Against Fascism* group; a staunch opponent of the EDL. Accounts from the USA are also present, acting as a further reminder of the international relationships within the counter-Jihad movement [31]. The final two small communities appear to consist of soccer fans, and are associated with Rangers Football Club from Scotland, and London-based soccer clubs respectively. The latter also contains a number of accounts affiliated with the EDL.

5.5   Case Study: German language

An interactions network (Fig. 5) was created for the German language data set, consisting of 208 nodes and 630 edges. As before, we measured the mean stability score ($\tau = 0.5$) and number of communities found for values of P in $[0.1, 0.9]$. The corresponding plots found in Fig. 6 show that the highest mean score was generated for $P = 0.9$, while the number of communities found had also stabilized at ~ 15. Five communities having at least ten members were found for $P = 0.9$ (74% of total network account nodes), and their corresponding hashtag

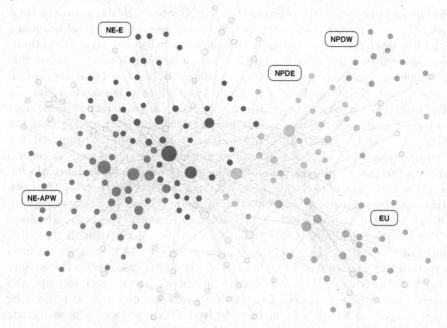

Fig. 5. German language interactions network with 208 nodes and 630 edges. Node size is proportional to degree. Consensus communities ($P = 0.9$, #members ≥ 10) are labelled.

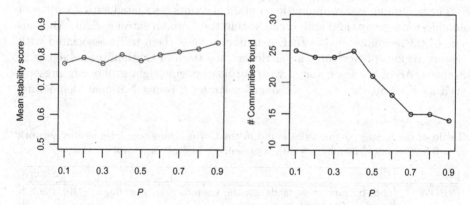

Fig. 6. German language mean stability scores (left) and number of communities found (right) for P in $[0.1, 0.9]$

descriptions, stability scores and manually annotated identifiers can be found in Table 4. Topic analysis was also performed with NMF, with number of topics $T = 10$, as lower values of T led to topics associated with relatively few account documents.

The first community consists of accounts associated with the *Nationaldemokr-atische Partei Deutschlands - National Democratic Party of Germany* (NPD) that appear to be localized to western regions of Germany. Hashtags such as *land-tagswahl/ltw...* (regional election) and *flugblatt* (flyer/leaflet/pamphlet) along with analysis of the tweet content may indicate mobilization prior to elections. Two topics most closely associated with this community include a Germany-wide NPD topic in addition to a general topic appearing to be related to street demonstrations. For example, *stolberg* refers to the town where a German teenager was killed by non-Germans in 2008, which has been the focus of annual extreme right commemorations. The EU community appears to be somewhat analogous to the NO2EU English language community, in that its membership is composed of a mixture of moderate and more extreme nationalist accounts bound by general opposition to the EU and related entities such as the European Stability Mechanism (*stopesm*), in addition to domestic German political parties (*cdu*, *fdp*, *spd*). Other notable members include a number of counter-Jihad accounts, along with others associated with relatively high-profile external media and blog websites.

The NPDE community is the second that can be associated with the NPD, which, in contrast to the NPDW community, appears to be largely localized to eastern German regions. In addition to NPD politicians, various Freies Netz (neo-Nazi collectives - FN) and "information/news portal" accounts are also present [33]. Similar variants of the *landtagswahl* hashtag and corresponding tweets to those of NPDW can also be observed. Other issues of interest include references to Thilo Sarrazin, a German politician and former Bundesbank executive who has criticized German immigration policy and proposed the abolition of the euro currency. As might be expected, the general NPD topic is linked to this community, while the anti-EU/political system topic is also prominent.

The remaining two communities contain accounts associated with a variety of non-electoral groups and individuals within the German extreme right. An analysis of the accounts in the NE-E community finds them to be associated with eastern regions of Germany, in particular, the town of Geithain near Leipzig in Sachsen. Accounts assocatiated with various extreme right groups are present, such as FN and the *Junge Nationaldemokraten* (Young National Democrats,

Table 4. Consensus communities found in the German language interactions network ($\tau = 0.5$, $P = 0.9$, #members ≥ 10), representing 74% of all network nodes

Id	Description	Size	Score
NPDW	npd, bochum, nrw, landtagswahl, wattenscheid, ovg, flugblatt, krefeld, anklage, bamberg	10	0.92
EU	euro, esm, stopesm, spd, islam, cdu, piraten, fdp, berlin, griechenland	26	0.82
NPDE	npd, ltwlsa, deutschland, ltw2011, berlin, linke, lsa, ltwmv, guttenberg, sarrazin	28	0.80
NE-E	geithain, apw, emwall, widerstand, jena, unsterblichen, gera, leipzig, volkstod, tdi	44	0.71
NE-APW	apw, gema, denkdran, volkstod, demokraten, dresden, hannover, spreelichter, 130abschaffen, unsterblichen	45	0.64

youth wing of the NPD). References to *Aktionsbüros* (coordination of activist activities) are also made. The *emwall* hashtag was popularly used during the 2012 UEFA European Football Championship, and accounts in this community used it to promote tweets suggesting that players having non-German ancestry should be excluded from the national squad. Separately, the *unsterblichen* (immortals) hashtag refers to anti-democratic flashmob marches that previously occurred sporadically throughout Germany in 2011 and 2012. These protests were linked to *Spreelichter*, an organization that was banned by the German authorities in 2012, whose account was also a member of this community [34]. They used social media to propagate national socialist-related material, including professional-quality videos of the marches themselves. In general, the accounts in this community appear to be quite active, with many tweets containing URLs linking to content hosted on external platforms such as YouTube or other dedicated websites.

Table 5. German language communities and associated NMF topics ($T = 10$, hashtag description cosine similarity ≥ 0.1)

Community	Similarity	Top 10 Topic Terms
NPDW	0.35	npd, landtagswahl, ltwlsa, ltwmv, nrw, sachsenanhalt, bochum, linke, wahlen, wattenscheid
	0.14	stolberg, dortmund, nrw, demonstration, demo, aachen, koln, brd, rheinland, munster
EU	0.39	esm, euro, stopesm, spd, cdu, deutschland, fdp, piraten, stoppesm, islam
NPDE	0.26	npd, landtagswahl, ltwlsa, ltwmv, nrw, sachsenanhalt, bochum, linke, wahlen, wattenscheid
	0.14	esm, euro, stopesm, spd, cdu, deutschland, fdp, piraten, stoppesm, islam
NE-E	0.17	emwall, geithain, tdi, leipzig, imc, linksunten, arbeiterkampftag, lvz, hof, dd2012
	0.16	widerstand, unsterblichen, altermedia, israel, wuppertal, hannover, dieunsterblichen, repression, abschiebar, spreelichter
	0.15	jena, raz10, gera, thuringen, volkstod, kahla, altenburg, demokraten, dresden, apw
	0.13	apw, volkstod, hannover, heldengedenken, demokratie, spreelichter, cottbus, guttenberg, dresden, vds
	0.10	stolberg, dortmund, nrw, demonstration, demo, aachen, koln, brd, rheinland, munster
NE-APW	0.40	apw, volkstod, hannover, heldengedenken, demokratie, spreelichter, cottbus, guttenberg, dresden, vds
	0.19	gema, denkdran, chemnitz, 5maerz, dresden, magdeburg, 13februar, 130abschaffen, demokraten, akt
	0.14	widerstand, unsterblichen, altermedia, israel, wuppertal, hannover, dieunsterblichen, repression, abschiebar, spreelichter
	0.13	jena, raz10, gera, thuringen, volkstod, kahla, altenburg, demokraten, dresden, apw

The second non-electoral community, NE-APW, appears to contain accounts from regions throughout Germany. The concept of *apw* (außerparlamentarischer Widerstand - non-parliamentary resistance) is prominent here, referring to actions taken outside of the democratic process. Other relevant hashtags include *13februar*, *denkdran*, *dresden* and *gema*, which all refer to the bombing of Dresden which began on February 13, 1945. The anniversary of this event is usually commemorated by extreme right groups each year. Also relevant is *volkstod*, which refers to the perceived destruction of the German race and traditions since World War II. This concept has separately featured in material distributed by alleged supporters of the National Socialist Underground [35], a group linked to a series of murders throughout Germany. The official account of the *Besseres Hannover* organization is present here; an initial ban by the German authorities was followed by this account being blocked by Twitter within Germany [36,37]. Both NE communities are associated with a variety of topics, perhaps reflecting the fragmented nature of non-electoral groups in the German extreme right. Relevant themes include street demonstrations (*stolberg*), some form of resistance (*apw*, *widerstand*) and media references such as *altermedia* (a collective of politically-incorrect/nationalist-oriented news websites).

5.6   Case Studies Discussion

For both the English and German language case studies, identifiable communities of accounts and related topics are clearly observable. When creating the original data sets, we purposely focused on non-electoral accounts and excluded electoral accounts affiliated with political parties. However, in both cases, the expansion of the data sets using reciprocal follower relationships resulted in the inclusion of electoral accounts, such as those affiliated with UKIP or the NPD. This potentially indicates the presence of some form of relationship between these two categories, if only at a passive level. Other similarities include the presence of distinct anti-EU communities and topics, in addition to the use of media accounts such as those associated with extreme right news websites and radio stations, along with external websites hosting media content. While geographical proximity is evident in most communities, linguistic proximity is a key factor in the existence of international connections such as those between certain counter-Jihad groups and individuals. Although the underlying ideology of certain communities can often be identified, this is less clear in other cases, especially when the mappings between communities and their associated topics are considered. Here, multiple topic mappings suggest a complexity and diversity in both the membership composition and interests of the corresponding community. However, this may also be related to data incompleteness, variances in Twitter usage patterns between different countries, and the fact that an opinion that may be legally voiced in one country may not be permitted in another.

We should mention that this sample of accounts does not provide full coverage of all extreme right Twitter activity, and the accounts and subsequent communities and topics are greatly dependent on the initial selection of relevant accounts. In this domain, the random sampling of Twitter accounts is unlikely to yield a representative data set, as it is probably safe to assume that the total number of extreme right accounts merely constitutes a small percentage of all accounts. The fact that we did not rely on hashtags for data selection, coupled with the expansion using reciprocal follower relationships goes some way to avoid "selecting on the dependent variable" [38]. However, as suggested by Boyd and Crawford, it is important to acknowledge all known data set limitations [39]. At the same time, the authors also recognize the value of small data sets, where research insights can be found at any level. In this case, in spite of data sampling and coverage issues, it is still possible to detect the presence of extreme right communities and topics on Twitter. In addition, they emphasize the importance of results interpretation, which we have addressed here in our discussion of the communities and topics. In relation to this, care should be taken when inferring conclusions from results. It is unclear as to how representative they may be of offline extreme right network activity. It may be the case that social media platforms are merely used by these groups to disseminate related material to a wider audience, with the majority of subsequent interaction occurring elsewhere, but it is naturally difficult to quantify the extent to which this occurs. Separately, Boyd and Crawford also raise the issue of ethics in relation to publicly accessible data, emphasising the need for accountability on the part of researchers. Here, we address this by restricting discussion to known extreme right groups and their affiliates without identifying any individual accounts.

6   Conclusions and Future Work

Extreme right groups have become increasingly active in social media platforms such as Twitter in recent years. We have presented an analysis of the activity of a selection of such groups using network representations based on reciprocal follower and interaction activity, in addition to topic analysis of their corresponding tweets. The existence of stable communities and associated topics within local interaction networks has been demonstrated, and we have also identified international relationships between groups across geopolitical boundaries. Although a certain awareness exists between accounts based on follower relationships, it would appear that mentions and retweets interactions indicate stronger relationships where linguistic and geographical proximity are highly influential, in particular, the use of the English language. In relation to this, media accounts such as those associated with extreme right news websites and radio stations, along with external websites hosting content such as music or video, are a popular mechanism for the dissemination of associated material.

In future work, we will address the issues of sampling and incompleteness in the data sets, where the emergence of new extreme right groups should also be considered. The temporal properties of these networks will also be studied to provide insight into the evolution of extreme right communities over time. Separately, we plan to investigate the use of probabilistic topic models that support the discovery of more specific topics similar to those found in this analysis.

Acknowledgements. This research was supported by 2CENTRE, the EU funded Cybercrime Centres of Excellence Network and Science Foundation Ireland Grant 08/SRC/I1407 (Clique: Graph and Network Analysis Cluster).

References

1. Hoffman, D.S.: The Web of Hate: Extremists Exploit the Internet. Anti-Defamation League (1996)
2. Burris, V., Smith, E., Strahm, A.: White Supremacist Networks on the Internet. Sociological Focus 33(2), 215–235 (2000)
3. Bartlett, J., Birdwell, J., Littler, M.: The New Face of Digital Populism. Demos (2011)
4. Baldauf, J., Groß, A., Rafael, S., Wolf, J.: Zwischen Propaganda und Mimikry. Neonazi-Strategien in Sozialen Netzwerken. Amadeu Antonio Stiftung (2011)
5. Chau, M., Xu, J.: Mining communities and their relationships in blogs: A study of online hate groups. Int. J. Hum.-Comput. Stud. 65(1), 57–70 (2007)
6. Tateo, L.: The Italian Extreme Right On-line Network: An Exploratory Study Using an Integrated Social Network Analysis and Content Analysis Approach. Journal of Computer-Mediated Communication 10(2) (2005)
7. Caiani, M., Wagemann, C.: Online Networks of the Italian and German Extreme Right. Information, Communication & Society 12(1), 66–109 (2009)
8. Zuev, D.: The movement against illegal immigration: analysis of the central node in the Russian extreme-right movement. Nations and Nationalism 16(2), 261–284 (2010)
9. Blee, K.M., Creasap, K.A.: Conservative and Right-Wing Movements. Annual Review of Sociology 36, 269–286 (2010)
10. Bermingham, A., Conway, M., McInerney, L., O'Hare, N., Smeaton, A.F.: Combining Social Network Analysis and Sentiment Analysis to Explore the Potential for Online Radicalisation. In: Proc. International Conference on Advances in Social Network Analysis and Mining, pp. 231–236. IEEE Computer Society (2009)
11. Sureka, A., Kumaraguru, P.: Mining YouTube to Discover Extremist Videos. Information Retrieval, 13–24 (2010)
12. Goodwin, M., Ramalingam, V.: The New Radical Right: Violent and Non-Violent Movements in Europe. Institute for Strategic Dialogue (2012)
13. Macskassy, S.: On the Study of Social Interactions in Twitter. In: Sixth International AAAI Conference on Weblogs and Social Media, ICWSM (2012)
14. Takhteyev, Y., Gruzd, A., Wellman, B.: Geography of Twitter networks. Social Networks 34(1), 73–81 (2012)
15. Kulshrestha, J., Kooti, F., Nikravesh, A., Gummadi, K.: Geographic Dissection of the Twitter Network. In: Proc. 6th International AAAI Conference on Weblogs and Social Media (2012)
16. Kwak, H., Lee, C., Park, H., Moon, S.: What is Twitter, a social network or a news media? In: Proceedings of the 19th International Conference on World Wide Web, WWW 2010, pp. 591–600. ACM, New York (2010)
17. Lancichinetti, A., Fortunato, S.: Consensus clustering in complex networks. Sci. Rep. 2 (March 2012)
18. Greene, D., O'Callaghan, D., Cunningham, P.: Identifying Topical Twitter Communities via User List Aggregation. In: Proc. 2nd International Workshop on Mining Communities and People Recommenders, COMMPER 2012 (2012)

19. Lancichinetti, A., Radicchi, F., Ramasco, J., Fortunato, S., Ben-Jacob, E.: Finding statistically significant communities in networks. PLoS ONE 6(4), e18961 (2011)
20. Hubert, L., Arabie, P.: Comparing partitions. Journal of Classification, 193–218 (1985)
21. Hannon, J., Bennett, M., Smyth, B.: Recommending Twitter users to follow using content and collaborative filtering approaches. In: Proc. 4th ACM Conference on Recommender Systems, RecSys 2010, pp. 199–206 (2010)
22. Steyvers, M., Griffiths, T.: Probabilistic Topic Models. In: Landauer, T., Mcnamara, D., Dennis, S., Kintsch, W. (eds.) Latent Semantic Analysis: A Road to Meaning, Laurence Erlbaum (2006)
23. Deerwester, S.C., Dumais, S.T., Landauer, T.K., Furnas, G.W., Harshman, R.A.: Indexing by latent semantic analysis. Journal of the American Society of Information Science 41(6), 391–407 (1990)
24. Blei, D.M., Ng, A.Y., Jordan, M.I.: Latent dirichlet allocation. J. Mach. Learn. Res. 3, 993–1022 (2003)
25. Lee, D.D., Seung, H.S.: Learning the parts of objects by non-negative matrix factorization. Nature 401, 788–791 (1999)
26. Ramage, D., Dumais, S., Liebling, D.: Characterizing microblogs with topic models. In: Proc. 4th International AAAI Conference on Weblogs and Social Media (2010)
27. Weng, J., Lim, E.P., Jiang, J., He, Q.: TwitterRank: finding topic-sensitive influential twitterers. In: Proc. 3rd ACM International Conference on Web Search and Data Mining, pp. 261–270. ACM (2010)
28. Saha, A., Sindhwani, V.: Learning Evolving and Emerging Topics in Social Media: A Dynamic NMF approach with Temporal Regularization. In: Proc. 5th ACM International Conference on Web Search and Data Mining, pp. 693–702. ACM (2012)
29. Chemudugunta, C., Smyth, P., Steyvers, M.: Modeling General and Specific Aspects of Documents with a Probabilistic Topic Model. In: Advances in Neural Information Processing Systems, pp. 241–248 (2006)
30. Boutsidis, C., Gallopoulos, E.: SVD based initialization: A head start for nonnegative matrix factorization. Pattern Recogn. 41(4), 1350–1362 (2008)
31. Goodwin, M.: The Roots of Extremism: The English Defence League and the Counter-Jihad Challenge. Chatham House (2013)
32. Bartlett, J., Littler, M.: Inside the EDL: Populist Politics in a Digital Age. Demos (2011)
33. Netz Gegen Nazis: Nach den Portal-Sperren im Internet: Hier geht der Hass weiter. netz-gegen-nazis.de (2012)
34. Radke, J.: Das Ende der Nazi-Masken-Show. Die Zeit (June 2012)
35. Fuchs, C., Müller, D.: NSU-Prozess: Die weißen Brüder. Die Zeit (April 2013)
36. Störungsmelder: Nach Razzia bei Neonazi-Gruppierung "Besseres Hannover" folgt ein Verbot. Die Zeit (September 2012)
37. Connolly, K.: Twitter blocks neo-Nazi account in Germany. The Guardian (October 2012)
38. Tufekci, Z.: Big Data: Pitfalls, Methods and Concepts for an Emergent Field. SSRN (March 2013)
39. Boyd, D., Crawford, K.: Critical questions for big data. Information, Communication & Society 15(5), 662–679 (2012)

Who will Interact with Whom?
A Case-Study in Second Life Using Online Social Network and Location-Based Social Network Features to Predict Interactions between Users

Michael Steurer[1] and Christoph Trattner[2]

[1] IICM, Graz University of Technology
Inffeldgasse 16c
msteurer@iicm.edu
[2] Know-Center, Graz University of Technology
Inffeldgasse 13/5
ctrattner@know-center.at

Abstract. Although considerable amount of work has been conducted recently of how to predict links between users in online social media, studies inducing features from different domain data are rare. In this paper we present the latest results of a project that studies the extent to which interactions – in our case directed and bi-directed message communication – between users in online social networks can be predicted by looking at features obtained from online and location-based social network data. To that end, we conducted a number of experiments on data obtained from the virtual world of Second Life. As our results reveal, location-based social network features outperform online social network features if we try to predict interactions between users. However, if we try to predict whether or not this communication was also reciprocal, we find that online social network features seem to be superior.

Keywords: online social networks, location-based social networks, link prediction problem, predicting interactions, predicting reciprocity, virtual worlds, Second Life.

1 Introduction

As a part of the recent hype in social network research, a high amount of attention and research activity was devoted to the problem of predicting links between users [17], *e.g.* the issue of forecasting whether or not two users u and v of a given online social network $G\langle V, E\rangle$ will interact with each other in the future. While considerable amount of work has been recently conducted of how to predict links between users in online social media, studies comparing different sources of knowledge are rare.

To contribute to this research, we present in this paper the latest results of a research project that aims to study the extent to which interactions – in our

M. Atzmueller et al. (Eds.): MUSE/MSM 2012, LNAI 8329, pp. 108–127, 2013.
© Springer-Verlag Berlin Heidelberg 2013

case directed and bi-directed message communications – in online social networks can be predicted inducing features from online social network and location-based social network data. To tackle this issue we trained a binary classifier that learned the relations between users u and v based on a number of features induced from online social network and location-based social network data. For the purpose of our study we furthermore differentiated between two types of feature sets – network topological features and homophilic features [22]. Since it is nearly impossible to obtain rich large-scale real-world online social and location-based data, our investigation focused on the virtual world of Second Life, where we could easily find and mine both sources of data. We obtained data from a resource called *My Second Life* which is a large-scale online social network for residents of Second Life. This social network can be compared to Facebook but aims at a different target group: residents of Second Life who interact with each other by sharing text messages, comments, and loves. Additionally, we were able to collect location-based social network data of residents in the virtual world by implementing the so-called in-world bots.

Overall, it is our interest to answer the following research questions:

- *RQ1:* To what extent do user pairs – interacting or not interacting with each other – differ based on social proximity features induced from the online social network and the location-based social network?

- *RQ2:* To what extent can we predict interactions between users and reciprocity of these interactions inducing features from both domains?

- *RQ3:* Which feature set (homophilic or topological) is most suitable to predict interactions between users and the reciprocity of these interactions?

To that end, we conducted a number of experiments using statistical methods and supervised learning approaches. As our statistical analysis reveals, there are many significant differences between user pairs with interactions and user pairs without interactions. For instance, users with interactions on the online social network have a shorter average distance between them in the location-based social network. To predict these interactions with supervised learning, we find that location-based social network features outperform online social network features to a great extent. However, if we try to predict reciprocal message communication between users, online social network features seem to be superior. Finally, we find that there are no clear patterns whether or not homophilic or network topological features perform better to predict interactions or reciprocity between users.

The paper is structured as follows: In Section 2, we discuss related work. In Section 3 we shortly introduce the dataset used for our experiments. In Section 4 we outline the set of features used for our experiments in Section 5. Section 6 presents the results of our study. Finally, Section 7 discusses our findings and concludes the paper.

Fig. 1. Sample of a user profile in the online social network *My Second Life*. Users can *post* text message on their walls or can communicate with each other by *commenting* or *loving* onto each other's posts.

2   Related Work

Although considerable amount of work has been recently conducted of how to predict links between users in online social media, studies exploiting different kinds of knowledge sources for the link prediction problem are rare. An example is a study conducted by Cranshaw *et al.* where the authors collected location data and Facebook friendship data through a mobile app [6]. Based on a number of experiments they show that the so-called place-entropy features are best suited to predict friendship between users in Facebook. Interestingly and contrary to our study, Cranshaw *et al.* only looked at the mobile side, i.e. they did not investigate features induced directly from the social network. Furthermore, they only considered friendship links and did not look at communication links as we do in our study. Another related work in this context are the studies of Guy *et al.* [11], [12], [10] where the authors investigate the similarity between users exploiting 9 different sources of data classified into three different classes: *people, things,* and *places.* Looking at only semantic features such as tags, they find that the so-called "tagged-with" feature performs well in all three different data category sources.

Probably one of the first projects investigating the link prediction problem from the network topological perspective in the context of online social media is a work conducted by Golder and Yardi [8]. In their paper they study the micro-blogging service Twitter and find "that two structural characteristics, transitivity and mutuality, are significant predictors of the desire to form new ties". The first paper investigating the extent to which reciprocity could be predicted in the online social media is a recent paper by Cheng *et al.* [4]. By applying a rich set of network based features including link prediction features from [17], they show that the so-called out-degree measure of a user in Twitter is the best feature to predict reciprocity. Another interesting work in this context is a study conducted by Yin *et al.* [23]. In their paper they investigate the link prediction problem within the micro-blogging system Twitter. The main contribution, apart from studying the performance of well established link prediction methods, is the introduction of a "novel personalized structure-based link prediction model" which "noticeably outperforms the state-of-the-art" methods. The first work studying the computational efficiency of network topological features in the online domain is a paper written by Fire *et al.* [7]. In their work they apply a rich set of over 20 features on a set of 5 different online social network sites with respect to their computational efficiency. Their study reveals that the so-called friends measure shows a good trade-off between accuracy and computational efficiency.

Another study in this context is a recent study conducted by Rowe *et al.* In their work [19] they study the link prediction problem, or the question who will follow whom, in the micro-blogging system *Tencent Weibo*. Looking at both – semantic and network topological features – they show that the predictability of links can be significantly improved by training a classifier that uses both. Although the work of Rowe *et al.* has considerable amount of overlap with our own work, their study only looked at features which could be directly induced from the online media site Tencent Weibo. Hence, contrary to our own work they did not include external knowledge such as location-based social network data as we do in our study. Finally, the last study to be mentioned is a work conducted by Scellato *et al.* [20]. Similar to our work they tried to exploit features from the location-based social network of *Gowalla* to predict links between users. However, in contrast to our work, they only focused on location-based social data and did not combine online social network and location-based social network data as we do in this paper. In their analysis over a period of three months they found that most of the links are formed between users that visit the same places or places that share similar properties.

3   Datasets

As stated in the introductory part of this paper we conducted our experiments on two types of datasets – online social network and location-based social data – both obtained from the virtual world of Second Life. The reasons for choosing Second Life over other real world sources are manifold: First, in contrast to networks such as Facebook, the online social network *My Second Life* does not restrict extensive crawling of user profiles. Second and contrary to real world online social networks, most profiles in *My Second Life* are public, i.e. we can mine a large fraction of the network. Third, in virtual worlds the location information of users can be harvested in an automated way whereas it is nearly impossible to obtain large-scale tracking data of users in the real world. In this section we describe the collection process for the data as used in our experiments.

3.1   Location-Based Social Network Dataset

The collection of the location-based social network dataset in Second Life was a two stage process: First a list of popular locations from the Second Life Event calendar[1] was crawled. Second, overall 15 in-world agents, the so-called in-world-bots, were implemented to teleport to these locations and gather location information of the users at a specific place.

In detail the procedure was the following: In order to harvest all events in Second Life we implemented a Web-crawler that runs on a daily bases to obtain all publicly announced events on the Second Life Event calendar. Allover, we were able to obtain data of 218,245 unique events during a period of ten months starting in March 2012.

In order to collect location data of the users we implemented overall 15 in-world agents on the basis of the open source command-line client *libopen-metaverse*[2]. Due to the modularity of the tool, we were able to enhance the functionality of our agents to teleport around in the virtual world to collect location data of all surrounding users in a region. This location information comprised the current region, x and y coordinates of the location within this region, and a time stamp. The pool of agents was controlled by a centralized instance sending our in-world bots to ongoing events. Due to the large amount of concurrent events in several regions of Second Life and the constraint that a bot was only able to obtain data of one single region at the same time, our sampling rate was set to a limit of 15 minutes. All in all, we were able to obtain over 13 Million data samples of 190,160 unique users visiting events with this kind of approach [21].

3.2   Online Social Network Dataset

In July 2011 Linden Labs introduced an online social network called *My Second Life*[3] similar to other social networks such as Google+ or Facebook. Residents of the virtual world can log-in with their in-world credentials, access their friend lists and have a so-called *Feed* that can be compared to the Google+ Stream or the Facebook Wall. The social interaction with other users is done by sharing text messages, screenshots, comments and so-called loves which can be seen equally to a Like on Facebook or a Plus in Google+ (see Figure 1). Furthermore, users can enhance their profiles by adding personal information such as interests, groups, etc.

We attempted to download the profile data of all 190,160 users found by the avatar-bots. In the next step we parsed the interaction-partners of these users and downloaded the profile information of the missing ones. This procedure was repeated until no new users could be found by our crawler. Finally, this procedure yielded in a dataset of 311,959 users with 300,657 of them opened to the public, and 135,181 with interactions on their feed.

---

[1] http://secondlife.com/community/events/
[2] http://lib.openmetaverse.org/
[3] https://my.secondlife.com/

Table 1. Basic metrics of the two networks and their combination used for the experiments

Name	Location-based G_M	Online G_F	$G_{FM} = G_F + G_M$
Type	undirected	directed	directed
Nodes	131,349	135,181	37,118
Edges	2,343,683	209,653	1,043,172
Degree	35.7	3.1	56.2

4   Feature Sets

As already outlined, it is our interest to predict interactions between users in online social networks based on features induced from online social network and location-based social network data. To that end, we induced two different types of feature sets from our data sources: network topological and homophilic features [22]. In order to start with the description of the different features calculated for our experiments we first describe the networks derived from the collected data.

The first network, referred to as *online social network*, was based on data obtained from the users profile where every edge in this directed network indicates communication between two users. This yielded in a network with 135,181 users and 209,653 edges. The second network, referred to as *location-based social network*, was based on the users location data where every edge in this undirected network indicated that two users were seen concurrently in the same region on two different days. This yielded in a network with 142,507 nodes and 3,773,316 edges. A summery of both networks can be found in Table 1. Figure 2 shows the degree distribution of the social network and location-based social network. Both networks show power-law qualities with an alpha of 1.55 and a corresponding fitting error of 0.13 for the online social network and and alpha value of 2.67 and a fitting error of 0.16 for the location-based social network [5].

4.1   Online Social Network: Topological Features

In social networks such as Facebook or Google+ the friendship of users is based on a mutual agreement with reciprocal confirmation. In contrast to this, users of the online social network *My Second Life* can post onto each others' walls without this mutual agreement. Hence, as a consequence, we considered the social network as a directed graph $G_F \langle V_F, E_F \rangle$ with V_F representing the users and $e = (u, v) \in E_F$ if user u posted, commented, or liked something on the feed of user v.

We defined the set of the neighbors of a node $v \in G_F$ as $\Gamma(v) = \{u \mid (u, v) \in E_F$ or $(v, u) \in E_F\}$ and based on this definition of neighborhood we used the following topological features:

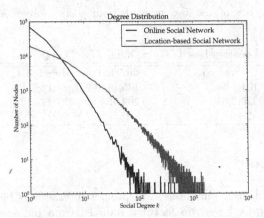

Fig. 2. Degree distributions for the online and the location-based social network

- *Common Neighbors* $F_{CN}(u, v)$. This represented number of interaction partners two users had in common.

$$F_{CN}(u, v) = |\Gamma(u) \cap \Gamma(v)|$$

For a directed network we split this into the number of common users $F_{CN}^+(u, v) = |\Gamma^+(u) \cap \Gamma^+(v)|$ to which both users sent messages to and the number of users $F_{CN}^-(u, v) = |\Gamma^-(u) \cap \Gamma^-(v)|$ from which both users received messages.

- *Jaccard's Coefficient* $F_{JC}(u, v)$. The ratio of the total number of neighbors and the number of common neighbors of two users was taken from [15] and is defined as follows.

$$F_{JC}(u, v) = \frac{|\Gamma(u) \cap \Gamma(v)|}{|\Gamma(u) \cup \Gamma(v)|}$$

For directed networks this could be split into two coefficients for received messages $F_{JC}^-(u, v) = \frac{|\Gamma^-(u) \cap \Gamma^-(v)|}{|\Gamma^-(u) \cup \Gamma^-(v)|}$ and sent messages $F_{JC}^+(u, v) = \frac{|\Gamma^+(u) \cap \Gamma^+(v)|}{|\Gamma^+(u) \cup \Gamma^+(v)|}$.

- *Adamic Adar* $F_{AA}(u, v)$. Instead of just counting the number of common neighbors with Jaccard's Coefficient in a network, this feature adds weights to all neighbors of a pair of users [1].

$$F_{AA}(u, v) = \sum_{z \in \Gamma(u) \cap \Gamma(v)} \frac{1}{log(|\Gamma(z)|)}$$

According to Cheng *et al.* this can be transformed into $F_{AA}^-(u, v) = \sum_{z \in \Gamma^-(u) \cap \Gamma^-(v)} \frac{1}{log(|\Gamma^-(z)|)}$ for directed networks [4].

- *Preferential Attachment Score $F_{PS}(u,v)$.* This feature took into account that active users, i.e. users with many interaction partners, are more likely to form new relationships than users with not so many interactions [2].

$$F_{PS}(u,v) = |\Gamma(u)| \cdot |\Gamma(v)|$$

The score was applied to a directed network with two different features: $F_{PS}^+(u,v) = |\Gamma^+(u)| \cdot |\Gamma^-(v)|$, respectively $F_{PS}^-(u,v) = |\Gamma^-(u)| \cdot |\Gamma^+(v)|$ [4].

4.2   Online Social Network: Homophilic Features

As stated before, users in Second Life can enhance their online social network profile by adding additional meta-data information such as interests or groups. As observed by a number of previous studies in this area [19], [22], homophily is an important variable in the context of the link prediction problem. We defined a set of homophilic features which we calculated as group and interest similarity between users u, v. Formally, we defined the groups of a user u as $\Delta(u)$, respectively her interests as $\Psi(u)$.

- *Common Groups $G_C(u,v)$.* The first feature we induce is the so-called common groups measure. It is calculated as follows.

$$G_C(u,v) = |\Delta(u) \cap \Delta(v)|$$

- *Jaccard's Coefficient for Groups $G_{JC}(u,v)$.* The second feature, is the so-called Jaccard's coefficient for groups. It was calculated in the following form.

$$G_{JC}(u,v) = \frac{|\Delta(u) \cap \Delta(v)|}{|\Delta(u) \cup \Delta(v)|}$$

- *Common Interests $I_C(u,v)$.* The third homophilic feature, was the number of interests two users had in common.

$$I_C(u,v) = |\Psi(u) \cap \Psi(v)|$$

- *Jaccard's Coefficient for Interests $I_{JC}(u,v)$.* And finally the last feature, which is a combination of total interests and common interests of the users.

$$I_{JC}(u,v) = \frac{|\Psi(u) \cap \Psi(v)|}{|\Psi(u) \cup \Psi(v)|}$$

4.3   Location-Based Social Network: Topological Features

We applied the same network topological feature calculations to the location-based social network as we did for the online social network. The network had edges between users that met on at least two days. Using this relation between in-world users defined the topological features similar to Section 4.1. Here, the neighbors of a node in the undirected location-based social network $G_M \langle V_M, E_M \rangle$ were defined as $\Theta(u) = \{v \mid (u,v) \in G_M\}$ and starting with this we defined the topological features as follows.

- *Common Neighbors* $M_{CN}(u, v)$.

$$M_{CN}(u, v) = |\Theta(u) \cap \Theta(v)|$$

- *Jaccard's Coefficient* $M_{JC}(u, v)$.

$$M_{JC}(u, v) = \frac{|\Theta(u) \cap \Theta(v)|}{|\Theta(u) \cup \Theta(v)|}$$

- *Adamic Adar* $M_{AA}(u, v)$.

$$M_{AA}(u, v) = \sum_{z \in \Theta(u) \cap \Theta(v)} \frac{1}{log(|\Theta(z)|)}$$

- *Preferential Attachment Score* $M_{PS}(u, v)$.

$$M_{PS}(u, v) = |\Theta(u)| \cdot |\Theta(v)|$$

4.4   Location-Based Social Network: Homophilic Features

These features were based on the actual distance between users, the regions they visit, and the number of days where they co-occurred concurrently. Let $O(u, v)$ be the co-locations of user u and user v, when both users resided in the same region concurrently. An observation $o \in O(u, v)$ was 4-tuple of region r, time stamp t, location coordinates of user u: $l_u = (x_u, y_u)$ and user v: $l_v = (x_v, y_v)$.

- *Physical Distance* $A_D(u, v)$. Whenever two users were observed concurrently, we measured the distance between them based on their x and y coordinates. As an indicator of their overall physical closeness, we therefore computed the average physical Euclidean distance between two users for all observations where both were present in the same region concurrently.

$$A_D(u, v) = \frac{1}{|O(u, v)|} \sum_{o \in O(u, v)} \|o(l_u) - o(l_v)\|$$

- *Days Seen* $A_S(u, v)$. This feature indicated the number of days when two users have been observed in the same region concurrently.

The regions of a user were defined as $P(u) = \{\rho \in P \mid$ user u was observed in region $P\}$ and so we computed the region properties of users as follows:

- *Common Regions* $R_C(u, v)$. The number of regions two users visited, not necessarily at the same time.

$$R_C(u, v) = |P(u) \cap P(v)|$$

- *Regions Seen Concurrently* $R_S(u, v)$. In contrast to the Common Regions feature, this feature took only the regions into account where both users were observed in the same region concurrently.

- *Observations Together* $R_O(u, v)$. This feature was taken from Cranshaw *et al.* [6] and represented the number of total regions of two users divided by the sum of each user's number of regions.

$$R_O(u, v) = \frac{|P_u \cup P_v|}{|P_u| + |P_v|}$$

5   Experimental Setup

All in all, we conducted two different experiments to study the extent to which interactions between users in online social networks can be predicted. Both experiments were based on the combination of the *online social network* G_F and the *location-based social network* G_M described in Section 4. To that end, we followed the approach of Guha *et al.* [9] in both experiments who suggest to create two datasets with an equal number of "positive edges" and "negative edges" for the binary classification problem. This results in balanced datasets for the test and training data and therefore in a baseline of 50% for the prediction when guessing randomly. For the evaluation of the binary classification problem we employed different supervised learning algorithms and used the area under the ROC curve (AUC) as our main evaluation metric to determine the performance of our features [14], [18]. We justified our findings with a 10-fold cross validation approach using the WEKA machine-learning suite [13].

In this section we describe in detail how the trainings and test data set for both experiments was generated.

5.1   Predicting Interactions

The task here is to predict whether or not two users interacted with each other on the feed by using topological and homophilic information of the online social network and the location-based social network. In the first step we computed the edge-features for the user-pairs as described in Section 4 for both networks independently. Then, in the second step we created the intersection of both networks as directed graph $G_{FM}\langle V_{FM}, E_{FM}\rangle$ where $V_{FM} = \{v|v \in V_F, v \in V_M\}$, and $E_{FM} = \{(u,v)|(u,v) \in E_M, (u,v) \in E_F, v \text{ and } u \in V_{FM}\}$. This newly created network consisted of 37,118 nodes and 1,014,352 pairs with location co-occurrences $((u,v) \in E_M)$, 36,213 pairs with social interaction $((u,v) \in E_F)$, and 7,393 edges with both $((u,v) \in E_M, E_F)$.

For the binary classification problem we uniformly selected 2,500 user-pairs with a social interaction and a location co-occurrence ("positive edges") $\{e^+ = (u,v)|e^+ \in E_{FM}, e^+ \in E_F, e^+ \in E_M\}$ and 2,500 user-pairs with a location co-occurrence but without a social interaction ("negative edges") $\{e^- = (u,v)| e^- \notin E_F, e^- \in E_M\}$. These edges, i.e. pairs of users, and the according edge features from both domains were used as datasets for all further evaluations and experiments.

Table 2. Means and standard errors of the features in the online social network and the location-based social network for the group of users having interactions with each other vs. the groups of users having no interactions (***=significant at level 0.001)

Features		Have Interactions	Have No Interactions
Online Social Network	Common Neighbors (in) $F_{CN}^-(u,v)$***	2.81 ± 0.32	0.02 ± 0.00
	Common Neighbors (out) $F_{CN}^+(u,v)$***	2.39 ± 0.27	0.01 ± 0.00
	Adamic Adar $F_{AA}(u,v)$***	14.65 ± 1.28	1.71 ± 0.18
	Jaccard's Coefficient (in) $F_{JC}^-(u,v)$***	0.05 ± 0.00	0.00 ± 0.00
	Jaccard's Coefficient (out) $F_{JC}^+(u,v)$***	0.04 ± 0.00	0.00 ± 0.00
	Preferential Attachment (in) $F_{PS}^-(u,v)$***	1566.55 ± 239.31	3.88 ± 0.64
	Preferential Attachment (out) $F_{PS}^+(u,v)$***	2088.94 ± 441.14	4.92 ± 1.53
	Common Groups $G_C(u,v)$***	1.92 ± 0.07	0.40 ± 0.02
	Jaccard's Coefficient $G_{JC}(u,v)$***	0.05 ± 0.00	0.01 ± 0.00
	Common Interests $I_C(u,v)$	0.07 ± 0.01	0.02 ± 0.00
	Jaccard's Coefficient $I_{JC}(u,v)$	0.00 ± 0.00	0.00 ± 0.00
Location-based Social Network	Common Neighbors $M_{CN}(u,v)$***	52.48 ± 4.98	83.61 ± 2.31
	Jaccard's Coefficient $M_{JC}(u,v)$***	0.20 ± 0.00	0.10 ± 0.00
	Preferential Attachment $M_{PS}(u,v)$***	218341.22 ± 164510.35	530640.88 ± 50352.29
	Adamic Adar $M_{AA}(u,v)$***	26.89 ± 3.19	36.43 ± 0.98
	Regions Seen $R_S(u,v)$***	2.81 ± 0.09	1.41 ± 0.02
	Common Regions $R_C(u,v)$***	3.59 ± 0.34	3.03 ± 0.08
	Observations Together $R_O(u,v)$***	0.22 ± 0.00	0.10 ± 0.00
	Distance $A_D(u,v)$***	10.32 ± 0.36	38.13 ± 0.95
	Days Seen $A_S(u,v)$***	7.34 ± 0.21	3.98 ± 0.09

5.2   Predicting Reciprocity

The task here is to predict whether two users had mutual activities on each other's walls, i.e. reciprocal interactions, by exploiting topological and homophilic information of the online social network and the location-based social network. We defined a reciprocal edge as $e'' = (u,v)|(u,v) \in G_F, (v,u) \in G_F$, a non-reciprocal edge as $e' = (u,v)|(u,v) \in G_F, (v,u) \notin G_F$, and used this difference for the binary classification problem. In contrast to the previous experiment we considered the online social network as undirected network for the computation of the edge-features but retained information about the reciprocity of the interactions. The edge features for the location-based social network were again considered as undirected. For the actual experiment we combined the online social network and the location-based social network to a new undirected network referred to as $G'_{FM}\langle V'_{FM}, E'_{FM}\rangle$ where $V'_{FM} = \{v|v \in V_F, v \in V_M\}$, and $E'_{FM} = \{(u,v)|(u,v) \in E_M, (u,v) \in E_F$ or $(v,u) \in E_F, v$ and $u \in V'_{FM}\}$. Out of the 7,393 user-pairs with a social interaction and a location co-occurrence we identified 1,431 reciprocal edges and 4,531 non-reciprocal edges in the online social network. For the binary classification task we uniformly selected pairs of users from the undirected network G'_{FM} with 1,000 reciprocal edges ("positive edges") and non-reciprocal edges ("negative edges") each. These edges, i.e. user-pairs with the according features, were used for all further evaluations and experiments.

Table 3. Means and standard errors of the features in the online social network and the location-based social network for the group of users having reciprocal interactions vs. the groups of users having no reciprocal interactions with each other (*=significant at level 0.1, **=significant at level 0.01, and ***=significant at level 0.001)

	Features	Reciprocal	Non Reciprocal
Online Social Network	Common Neighbors $F_{CN}(u,v)$***	10.20 ± 1.10	0.80 ± 0.10
	Adamic Adar $F_{AA}(u,v)$***	6.46 ± 0.61	0.71 ± 0.06
	Jaccard's Coefficient $F_{JC}(u,v)$***	0.08 ± 0.00	0.04 ± 0.00
	Preferential Attachment $F_{PS}(u,v)$***	12544.28 ± 2066.82	403.15 ± 93.73
	Common Groups $G_C(u,v)$	2.04 ± 0.11	1.81 ± 0.10
	Jaccard's Coefficient $G_{JC}(u,v)$	0.06 ± 0.00	0.05 ± 0.00
	Common Interests $I_C(u,v)$	0.12 ± 0.02	0.05 ± 0.01
	Jaccard's Coefficient $I_{JC}(u,v)$	0.01 ± 0.00	0.00 ± 0.00
Location-based Social Network	Common Neighbors $M_{CN}(u,v)$***	42.59 ± 2.67	61.29 ± 11.96
	Jaccard's Coefficient $M_{JC}(u,v)$**	0.2 ± 0.01	0.19 ± 0.01
	Preferential Attachment $M_{PS}(u,v)$*	41663.58 ± 4547.60	473151.99 ± 411215.48
	Adamic Adar $M_{AA}(u,v)$	21.01 ± 1.30	32.25 ± 7.79
	Regions Seen $R_S(u,v)$	2.82 ± 0.10	2.71 ± 0.18
	Common Regions $R_C(u,v)$	3.25 ± 0.12	4.00 ± 0.83
	Observations Together $R_O(u,v)$	0.23 ± 0.00	0.21 ± 0.00
	Distance $A_D(u,v)$**	9.35 ± 0.48	11.19 ± 0.57
	Days Seen $A_S(u,v)$**	7.22 ± 0.31	6.96 ± 0.33

6   Results

Before we start with the analysis of how to predict interactions between users, we show the differences between user pairs with and without interactions in the social network, respectively user pairs with reciprocal and non-reciprocal interactions for both domains. Both the Anderson-Darling test and the one-sampled Kolmogorov-Smirnov test showed that none of the distributions of the features described in Section 4 were normally distributed. Hence, and similar to Bischoff [3], we compared the variances of all features using a Levene test ($p < 0.01$). To test for significant differences of the means, we employed Mann-Whitney-Wilcoxon test in case of equal variances and a two-sided Kolmogorov-Smirnov test in case of unequal variances. The differences of the means between the groups of users regarding their interaction type can be found in Table 3 and 2. Overall, we found the following:

- *Interactions:* Mean values of topological features in the online social network were significantly higher for user pairs with interactions compared to users without interactions. For homophilic features, a significant difference between user pairs was observed for features based on group affiliation whereas features based on specified interests did not show significant differences. Topological features in the location-based social network also showed significant differences between users but contrary, users with no interactions had a higher number of common neighbors, preferential attachment score,

Table 4. Overall results AUC and (ACC) of the Logistic Regression learning approach for predicting interactions between users and their reciprocity in the online social network of Second Life using online social network and location-based social network features

	Feature Sets		Interaction	Reciprocity
Logistic Regression	Online Social Network	Topological	0.878 (71.8%)	0.676 (64.9%)
		Homophilic	0.640 (63.4%)	0.507 (52.5%)
		Combined	**0.863 (76.8%)**	**0.679 (64.8%)**
	Location-based Social Network	Topological	0.858 (76.7%)	0.530 (51.2%)
		Homophilic	0.885 (80.6%)	0.556 (54.4%)
		Combined	**0.919 (84.8%)**	**0.551 (53.5%)**
	All Features		**0.953 (89.6%)**	**0.709 (65.2%)**

and Adamic Adar score. Users with interactions had more common regions and observations, and they saw each other on more days. Furthermore, user pairs with interactions in the online social network had a significantly shorter average distance between them.

– *Reciprocity:* The differences between user pairs with reciprocal interactions and non-reciprocal interactions can be found in Table 3. The results revealed significant differences between users in the online social network for all topological features but no significant differences for homophilic features. Comparing differences between user pairs also showed significant differences in the topological features of the location-based social network (Common Neighbors, Jaccard's Coefficient and Preferential Attachment Score) but only the average distance between users and the number of days they saw each other was significantly different for the homophilic features

In the remainder of this section we present the results obtained from the two supervised learning experiments described in Section 5. As learning strategy we used the *Logistic Regression* learning algorithm since it can be easily implemented and interpreted [16].

6.1 Predicting Interactions: Online Social Network vs. Location-Based Social Network Features

The results of the first experiment can be found in Table 4 where we present the outcome of the prediction model for two different sources of knowledge and the according feature sets.

The values in the table represent the area under the ROC curve (AUC) and the accuracy of the prediction (ACC) as metrics for the predictability with a baseline for the binary classification problem of 0.5 AUC. As we can see, using topological features from the online social network improved the predictability of interactions between users by +37.8% whereas homophilic features (groups and interests) enhanced the baseline by +14.0%. In contrast to this, topological

Table 5. Coefficients of the Logistic Regression when all topological and homophilic features from both domains are used simultaneously in the dataset (***=significant at level 0.001)

	Features	Interactions	Reciprocity
Online Social Network	Common Neighbors (in) $F_{CN}^-(u,v)$	-1.782615***	–
	Common Neighbors (out) $F_{CN}^+(u,v)$	0.138448***	–
	Common Neighbors $F_{CN}(u,v)$	–	**-0.658291*****
	Adamic Adar $F_{AA}(u,v)$	0.196078	-0.108824***
	Jaccard's Coefficient (in) $F_{JC}^-(u,v)$	0.025060***	–
	Jaccard's Coefficient (out) $F_{JC}^+(u,v)$	2.416276 ***	–
	Jaccard's Coefficient $F_{JC}(u,v)$	–	0.495911***
	Preferential Attachment (in) $F_{PS}^-(u,v)$	**7.405495*****	–
	Preferential Attachment (out) $F_{PS}^+(u,v)$	-0.000097	–
	Preferential Attachment $F_{PS}(u,v)$	–	-1.107698
	Common Groups $G_C(u,v)$	-0.000066***	-0.000040***
	Jaccard's Coefficient $G_{JC}(u,v)$	0.216582***	-0.046399
	Common Interests $I_C(u,v)$	-1.230746	1.732937
	Jaccard's Coefficient $I_{JC}(u,v)$	0.932973	7.158616
Location-based Social Network	Common Neighbors $M_{CN}(u,v)$	-0.019859***	-0.004276
	Jaccard's Coefficient $M_{JC}(u,v)$	-0.001736***	-0.000470***
	Preferential Attachment $M_{PS}(u,v)$	0.000551***	0.000574
	Adamic Adar $M_{AA}(u,v)$	0.000001***	0.000000
	Regions Seen $R_S(u,v)$	0.294520	-0.101258
	Common Regions $R_C(u,v)$	0.717518***	0.093925
	Observations Together $R_O(u,v)$	0.022711***	-0.064381
	Distance $A_D(u,v)$	**10.570453*****	**1.158166*****
	Days Seen $A_S(u,v)$	-0.010596***	-0.002153

features from the location-based social network improved the baseline by +35.8% whereas homophilic features improved it by +38.5%. The combined topological and homophilic features from either networks resulted in a predictability of 0.953 AUC outperforming the baseline by +45.3%.

Overall, and interestingly, looking at the feature set in Table 4 we can see that location-based features were a great source to predict interactions between users in online social networks and they even outperformed online social network features. To evaluate the predictability of interactions of features separately, we present the coefficients of the Logistic Regression algorithm and their levels of significance when all features were used simultaneously. Table 5 shows that Preferential Attachment Score for incoming messages $F_{PS}^-(u,v)$ in the online social network and the average distance between users $A_D(u,v)$ in the location-based social network were most impacting features. Struck-out values in the table indicate a significance value of $p > 0.05$. To give an overview of the correlation of the features, we calculated the pair-wise Spearman-rank correlation of the used features from both domains as shown in Table 6.

Table 6. Spearman's Correlation Matrix (*=significant at level 0.1, **=significant at level 0.01, and ***=significant at level 0.001)

		F_{CN}^-	F_{CN}^+	F_{AA}	F_{JC}^-	F_{JC}^+	F_{PS}^-	F_{PS}^+	G_C	G_{JC}	I_C	L_{JC}	M_{CN}	M_{JC}	M_{PS}	M_{AA}	R_S	R_C	R_O	A_D	A_S
Online Social Network	F_{CN}^-	1.00																			
	F_{CN}^+	0.55***	1.00																		
	F_{AA}	0.49***	0.41***	1.00																	
	F_{JC}^-	0.99***	0.51***	0.47***	1.00																
	F_{JC}^+	0.53***	1.00***	0.40***	0.50***	1.00															
	F_{PS}^-	0.48***	0.47***	0.47***	0.46***	0.46***	1.00														
	F_{PS}^+	0.44***	0.49***	0.41***	0.47***	0.47***	0.30***	1.00													
	G_C	0.16***	0.56***	0.17***	0.09***	0.09***	0.19***	0.06***	1.00												
	G_{JC}	0.16***	0.08***	0.17***	0.17***	0.18***	0.06***	0.99***	0.03*	1.00											
	I_C	0.11***	0.13***	0.10***	0.12***	0.13***	0.12***	0.04*	0.04*	0.04*	1.00										
	L_{JC}	0.11***	0.12***	0.10***	0.12***	0.13***	0.12***	0.04*	0.06***	0.06***	1.00***	1.00									
Location-based Social Network	M_{CN}	0.13***	0.09***	0.08***	0.10***	0.10***	0.06***	0.20***	0.06***	0.20***	0.02	0.02	1.00								
	M_{JC}	0.03*	0.01	0.14***	0.01	0.05***	0.02	0.11***	0.01	0.10***	0.01	0.01	0.74***	1.00							
	M_{PS}	0.18***	0.14***	0.07***	0.15***	0.15***	0.07***	0.24***	0.02	0.26***	0.03*	0.03*	0.45***	0.32***	1.00						
	M_{AA}	-0.16***	-0.11***	-0.10***	-0.17***	-0.16***	-0.19***	0.20***	0.04*	0.20***	-0.04*	-0.04*	-0.19***	-0.19***	0.52***	1.00					
	R_S	-0.20***	-0.16***	-0.14***	-0.21***	-0.16***	-0.35***	-0.22***	-0.03*	-0.23***	-0.03*	-0.03*	-0.22***	0.10***	-0.53***	0.52***	1.00				
	R_C	-0.16***	-0.13***	-0.14***	-0.17***	-0.14***	-0.32***	-0.20***	-0.01	-0.20***	-0.03*	-0.03*	-0.05***	-0.05***	-0.51***	0.32***	0.94***	1.00			
	R_O	-0.18***	-0.14***	-0.12***	-0.15***	-0.24***	-0.18***	-0.19***	-0.03*	-0.19***	-0.03*	-0.03*	-0.18***	0.16***	-0.47***	0.50***	0.34***	0.78***	1.00		
	A_D	-0.07***	-0.05***	-0.06***	-0.07***	-0.09***	-0.02	-0.02	-0.03*	-0.03	-0.01	-0.01	-0.08***	0.14***	0.33***	0.61***	0.10***	0.14***	0.51***	1.00	
	A_S	0.14***	0.11***	0.14***	0.12***	0.16***	0.08***	0.20***	-0.03	0.20***	0.02	0.02	0.38***	0.31***	0.21***	-0.06***	-0.30***	0.23***	0.51***	0.51***	1.00

Online Social Network   Location-based Social Network

6.2 Predicting Reciprocity: Online Social Network vs. Location-Based Social Network Features

The results of the second experiment can be found in Table 4 where we present the area under the ROC curve (AUC) and the accuracy of the prediction (ACC). As in the previous experiment the baseline for randomly guessing is 0.5 AUC due to the balanced dataset.

Using topological features from the online social network increased the predictability of reciprocity by +17.6% whereas homophilic features alone (groups and interests) performed as badly as the baseline. Due to the little predictive power of the homophilic features the combination of all features in the online social network results in a prediction gain of +17.6% which is equal to topological features alone. In contrast to this, topological features from the location-based social network improved the baseline approach by +3.0% for the topological features and by +5.6% for the homophilic features. The combination of feature sets in the location-based social network boosted the predictability by +5.1%. The combination of features from either domains elevated the predictability of the reciprocity between two users up to 0.709 AUC, which is a boost of +20.9% if compared to the baseline of 0.5 AUC. Similar to the previous experiment, we computed the coefficients of the Logistic Regression algorithm in Table 5. In the online social network domain the Common Neighbors feature $F_{CN}(u, v)$ and in the location-based social network domain the distance between users $A_D(u, v)$ had the highest and most significant values. Again, struck-out values indicate a significance $p > 0.05$.

6.3 Verification of Stability: Predicting Interactions and Reciprocity with SVM and Random Forrest

The results of the conducted experiments based on LogisticRegression clearly showed that features from the location-based social network are better suited to predict interactions between users, whereas features from the online social network are better suited to predict reciprocity of interactions. However, to verify the stability of these findings we employed two additional learning algorithms: *Random Forest* and *Support Vector Machine* which are well suited for high dimensional, numeric and inter-dependent attributes (see Table 6) [3], [16]. The results of these learning algorithms are presented in Tables 7 and 8. Overall, the results can be interpreted as follows:

- *Predicting Interactions:* Using Logistic Regression, features from the location-based social network outperformed features from the online social network and similar results were observed for *Support Vector Machine* and *Random Forest*. In both cases features of the location-based social network resulted in a better prediction of interactions than features from the online social network. Overall, the performance of the combined feature set using Support Vector Machine was 0.882 AUC and using Random Forest was 0.979 AUC.

Table 7. Overall results AUC and (ACC) of the SVM learning approach for predicting interactions between users and their reciprocity in the online social network of Second Life using online social network and location-based social network features

	Feature Sets		Interactions	Reciprocity
SVM	Online Social Network	Topological	0.669 (66.9%)	0.646 (64.6%)
		Homophilic	0.638 (63.8%)	0.522 (52.2%)
		Combined	**0.737 (73.7%)**	**0.639 (63.9%)**
	Location-based Social Network	Topological	0.793 (79.3%)	0.529 (52.9%)
		Homophilic	0.761 (76.1%)	0.515 (51.5%)
		Combined	**0.849 (84.9%)**	**0.539 (53.9%)**
	All Features		**0.882 (88.2%)**	**0.638 (63.8%)**

Table 8. Overall results AUC and (ACC) of the Random Forrest learning approach for predicting interactions between users and their reciprocity in the online social network of Second Life using online social network and location-based social network features

	Feature Sets		Interactions	Reciprocity
Random Forest	Online Social Network	Topological	0.893 (79.7%)	0.628 (62.2%)
		Homophilic	0.624 (62,8%)	0.488 (50.4%)
		Combined	**0.910 (82.5%)**	**0.635 (60.5%)**
	Location-based Social Network	Topological	0.852 (77.9%)	0.530 (52.2%)
		Homophilic	0.872 (80.3%)	0.479 (49.2%)
		Combined	**0.916 (85.7%)**	**0.550 (53.2%)**
	All Features		**0.979 (93.0%)**	**0.684 (62.8%)**

- *Predicting Reciprocity:* For the prediction of reciprocity of interactions between users using Logistic Regression, online social network features outperformed location-based social network features. For other learning algorithms we found similar results as features from the online social network also outperformed features from the location-based social network. The combination of all features from both domains predicted reciprocity of interactions with 0.652 AUC using Support Vector Machine respectively 0.684 using Random Forest.

7   Discussion and Conclusions

In this work we harvested data from two Second Life related data sources: an online social network with text-based interactions and a location-based social network with position data. We modeled the social proximity between users with topological and homophilic network features and conducted two experiments.

To answer the first research question *RQ1*, we compared different features of user pairs regarding their interactions and the reciprocity of these interactions.

This analysis revealed that pairs with interactions were tighter connected in the online social network but the opposite was observed for the location-based social network. A possible explanation is that users in Second Life are allowed to directly "jump" to different regions in the whole virtual world but see the present users only upon arrival. We believe that users are more likely to stay in a region if they know present users, i.e. they have interactions on the online social network. This mobility activity could explain the tight connections in the location-based social network. This assumption is supported by homophilic features from both networks: users with interactions had more common groups, regions, and they saw each other on more days. Furthermore, the average distance was significantly shorter than users without interactions. All observed features were significantly different except interest based features but we assume this is due to the sparse data. The found results for predicting reciprocity of interactions was similar to the prediction of interactions themselves. User pairs with reciprocal interactions had tight connections in the online social network but the opposite was observed for the location-based social network. Again, homophilic features of user pairs with reciprocal interactions indicated a higher alikeness in both networks.

For the second research question $RQ2$ we predicted interactions and the reciprocity of these interactions. To do so, we chose Logistic Regression because it is easy to implement and interpret. We observed that interactions can be better predicted with features from the location-based social network than with features from the social network. Surprisingly, the opposite was observed for the reciprocity of interactions. In both experiments we found the combination of features from both networks outperforming either networks: Interactions could be predicted with 0.953 AUC and the reciprocity of these interactions with 0.709 AUC. The Logistic Regression coefficients of the features unveiled that a short average distance between users is a good indicator for interactions and their reciprocity. To verify our results that online social network features outperform features from the location-based social network for the prediction of interactions and vice versa for the prediction of reciprocity, we used two additional learning algorithms: Support Vector Machines and the Random Forest learning approach. Both algorithms approved the observations made in the experiment with Logistic Regression.

To answer the third research question $RQ3$, we compared homophilic features and topological features regarding the predictability of interactions and their reciprocity. Interestingly, we could not find a stable pattern over all experiments, as it was for instance proposed by Rowe et al. [19]. Although topological features of the online social network outperformed homophilic features in all three learning algorithms we found variation of the results for the location-based social network. Using Logistic Regression homophilic features performed better than topological features but in contrast, the opposite was observed for Support Vector Machines. With Random Forest homophilic features were better suited for the prediction of interactions but homophilic features were better suited for the reciprocity of interactions.

For future work, it is planned to dig deeper into the data and to address issues such as the variety of time (which we did not address in this study) or the issue why reciprocal links seem to be better predicted with social network features than with position data. Furthermore, we plan to extend our approach to predict other relations between users besides communicational interactions such as for instance partnership which can be also mined from the social network of Second Life. Finally, it is our interest to switch from supervised to unsupervised learning.

Acknowledgements. This work is supported by the Know-Center. The Know-Center is funded within the Austrian COMET Program - Competence Centers for Excellent Technologies - under the auspices of the Austrian Ministry of Transport, Innovation and Technology, the Austrian Ministry of Economics and Labor and by the State of Styria. COMET is managed by the Austrian Research Promotion Agency (FFG).

References

1. Adamic, L., Adar, E.: Friends and neighbors on the web. Social Networks 25(3), 211–230 (2003)
2. Barabási, A., Albert, R.: Emergence of scaling in random networks. Science 286(5439), 509–512 (1999)
3. Bischoff, K.: We love rock'n'roll: analyzing and predicting friendship links in Last. fm. In: Proceedings of the 3rd Annual ACM Web Science Conference, pp. 47–56. ACM (2012)
4. Cheng, J., Romero, D., Meeder, B., Kleinberg, J.: Predicting reciprocity in social networks. In: 2011 IEEE Third International Conference on Privacy, Security, Risk and Trust (Passat) and 2011 IEEE Third International Conference on Social Computing (Socialcom), pp. 49–56. IEEE (2011)
5. Clauset, A., Shalizi, C.R., Newman, M.: Power-law distributions in empirical data, arXiv preprint arXiv:0706.1062, 64 (2007)
6. Cranshaw, J., Toch, E., Hong, J., Kittur, A., Sadeh, N.: Bridging the gap between physical location and online social networks. In: Proceedings of the 12th ACM International Conference on Ubiquitous Computing, pp. 119–128. ACM (2010)
7. Fire, M., Tenenboim, L., Lesser, O., Puzis, R., Rokach, L., Elovici, Y.: Link prediction in social networks using computationally efficient topological features. In: 2011 IEEE Third International Conference on Privacy, Security, Risk and Trust (Passat) and 2011 IEEE Third International Conference on Social Computing (Socialcom), pp. 73–80. IEEE (2011)
8. Golder, S., Yardi, S.: Structural predictors of tie formation in twitter: Transitivity and mutuality. In: 2010 IEEE Second International Conference on Social Computing (SocialCom), pp. 88–95. IEEE (2010)
9. Guha, R., Kumar, R., Raghavan, P., Tomkins, A.: Propagation of trust and distrust. In: Proceedings of the 13th International Conference on World Wide Web, pp. 403–412. ACM (2004)
10. Guy, I., Jacovi, M., Perer, A., Ronen, I., Uziel, E.: Same places, same things, same people?: mining user similarity on social media. In: Proceedings of the 2010 ACM Conference on Computer Supported Cooperative Work, CSCW 2010, pp. 41–50. ACM, New York (2010)

11. Guy, I., Jacovi, M., Shahar, E., Meshulam, N., Soroka, V., Farrell, S.: Harvesting with sonar: the value of aggregating social network information. In: Proceedings of the SIGCHI Conference on Human Factors in Computing Systems, CHI 2008, pp. 1017–1026. ACM, New York (2008)

12. Guy, I., Zwerdling, N., Carmel, D., Ronen, I., Uziel, E., Yogev, S., Ofek-Koifman, S.: Personalized recommendation of social software items based on social relations. In: Proceedings of the Third ACM Conference on Recommender Systems, RecSys 2009, pp. 53–60. ACM, New York (2009)

13. Hall, M., Frank, E., Holmes, G., Pfahringer, B., Reutemann, P., Witten, I.: The weka data mining software: an update. ACM SIGKDD Explorations Newsletter 11(1), 10–18 (2009)

14. Huang, J., Ling, C.X.: Using auc and accuracy in evaluating learning algorithms. IEEE Trans. on Knowl. and Data Eng. 17(3), 299–310 (2005)

15. Jain, A., Dubes, R.: Algorithms for clustering data. Prentice-Hall, Inc. (1988)

16. Jones, J.J., Settle, J.E., Bond, R.M., Fariss, C.J., Marlow, C., Fowler, J.H.: Inferring tie strength from online directed behavior. PloS one 8(1), e52168 (2013)

17. Liben-Nowell, D., Kleinberg, J.: The link-prediction problem for social networks. Journal of the American Society for Information Science and Technology 58(7), 1019–1031 (2007)

18. Ling, C.X., Huang, J., Zhang, H.: Auc: a statistically consistent and more discriminating measure than accuracy. In: International Joint Conference on Artificial Intelligence, vol. 18, pp. 519–526. Lawrence Erlbaum Associates Ltd. (2003)

19. Rowe, M., Stankovic, M., Alani, H.: Who will follow whom? exploiting semantics for link prediction in attention-information networks (2012)

20. Scellato, S., Noulas, A., Mascolo, C.: Exploiting place features in link prediction on location-based social networks. In: Proceedings of the 17th ACM SIGKDD International Conference on Knowledge Discovery and Data Mining, pp. 1046–1054. ACM (2011)

21. Steurer, M., Trattner, C., Kappe, F.: Success factors of events in virtual worlds a case study in second life. In: NetGames, pp. 1–2 (2012)

22. Wang, D., Pedreschi, D., Song, C., Giannotti, F., Barabasi, A.: Human mobility, social ties, and link prediction. In: Proceedings of the 17th ACM SIGKDD International Conference on Knowledge Discovery and Data Mining, pp. 1100–1108. ACM (2011)

23. Yin, D., Hong, L., Davison, B.: Structural link analysis and prediction in microblogs. In: Proceedings of the 20th ACM International Conference on Information and Knowledge Management, pp. 1163–1168. ACM (2011)

Identifying Influential Users by Their Postings in Social Networks

Beiming Sun and Vincent TY Ng

Department of Computing,
Hong Kong Polytechnic University, Hong Kong
{csbsun,cstyng}@comp.polyu.edu.hk

Abstract. Much research effort has been conducted to analyze information of social networks, such as finding the influential users. Our aim is to identify the most influential users based on their interactions in posting on a given topic. We first proposes a graph model of online posts, which represents the relationships between online posts of one topic, so as to find the influential posts on the topic. Based on the influential posts found, the post graph is transformed to a user graph that can be used to discover influential users with improved influence measures. Finally the most influential users can be determined by considering the properties and measures from both graphs. In our work, two types of influences are defined based on two roles: starter and connecter. A starter is followed by many others, similar to a hub in a network; while a connecter is to help bridging two different starters and their corresponding clusters. In this paper, different measures on the graphs are introduced to calculate the influences on the two roles.

1   Introduction

With the rapid development and increased popularity of social networks, more and more interests have been made in obtaining information from social networking websites for analyzing people's behaviors. Our research is focusing on identifying the influential social network users; as it can help to increase the marketing efficiency, and also can be utilized to gather opinions and information on particular topics as well as to predict the trends. In order to find these influential users, the first problem is to measure a user's influence on social networks. In the past, there has been a lot of work on judging the influence of users on a specific social networking site. For example, many measurement metrics have made use of the relationship between users (i.e. follower / followee) in Twitter. However, they mostly ignore the interactions of users in their online posts. Moreover, without the consideration of the contents posted by users, they are not able to tell the influence of users on different topics.

In order to identify the influential users or leaders within a topic, we first obtain a measure of the influence of online posts on that topic. Next, we identify the most influential posts, and then based on their authors we further measure and compare the influence of users. In this paper, there are two types of influences based on the two roles: starter and connecter. A starter is followed by many others, similar to a hub in a network, so it should have certain influence. The connecter is also regarded to be

M. Atzmueller et al. (Eds.): MUSE/MSM 2012, LNAI 8329, pp. 128–151, 2013.
© Springer-Verlag Berlin Heidelberg 2013

influential when it links starters together. Both types are considered as influential in online posts as well as users.

The approach is to first figure out the relationship between online posts. Usually, posts are considered to be related in a thread or a chain. However, their relationships can be more complicated in certain cases. For example, a post is replying to a previous post while its content refers to a different one. Other than these direct responses as explicit relationships, there is also implicit relationship between online posts. For example, a user has read a post online. Instead of directly replying to it, he writes a new post on this topic. In this scenario, the two posts are considered to be implicitly related, because the action of later posting is influenced by the earlier one [1, 2]. Considering these situations, we proposed a graph model to represent the relationships between online posts on a topic. With the information of the explicit and implicit relationships between posts, the model tries to identify the most influential posts and users based on their direct interactions as well as the underlying relationships on the same topic. Three measurement methods are used to assess the influences of posts and to identify starters and connecters. Based on the influential posts found, we transformed the post graph to the user graph, and then refined the influence measures of users acting as starter and connecter. Finally, the most influential users can be identified by considering the properties and measures from both graphs.

The rest of the paper is organized as follows. Section 2 reviews some related works and a graph model is defined in section 3 to represent the relationships between online posts. After that, three different methods of influence measure are proposed based on the graph model. Section 5 defines the user graph model. The next session presents the conversion from the post graph to user graph, and the measurements of user influences. Section 7 discusses the tests with different cases to verify our models. Finally, we summarize the paper and suggest for future work in the last section.

2   Related Work

Many methods have been proposed to measure users' influence on Twitter. A popular metric of influence is the number of a user's follower [3]. It makes the assumption that all followers will read the contents published by that user. Yet, this method ignores the different ways for users to interact with the online contents. There are also many online tools to measure a user's influence on social network, such as Klout Score [4] and Twinfluence [5]. However, they cannot tell the influences of users on different topics. In [6], the TwitterRank algorithm, which is an extension of PageRank, was proposed to measure the user influence on Twitter taking both the topical similarity between users and the link structure into account. TunkRank [21] is another adaptation of PageRank. It makes the assumption that if a user reads a tweet from his friend he will retweet it with a constant probability. The influence is calculated recursively considering the attention a user can give to his friends, and that their followers could give to them. These methods do not consider users' interaction in posting. Yet, it is interesting to judge their influences not by their relations of friends in static structure, but based on the dynamic interaction in online contents.

As for the work of role detection on social media, Hansen et al. defined the social role of discussion starters based on graph metrics [7]. Discussion starters mostly receive messages often from people who are well-connected to each other, and they can be identified by low in-degree, high out-degree and high clustering coefficient in the graph. This metric does not suit our model, because the clustering coefficient is better to deal with an undirected graph or a directed graph with loops. Mathioudakis and Koudas did similar work [8] in distinguishing starters and followers on blogs. The starter does not mean the first one to open the discussion but the one who triggers an intense discussion. They expected that a blogger, who primarily generates posts that others link (inlinks) over a significant period of time, could be a starter, and the bloggers who primarily generate posts that links to other blog posts (outlinks) would be followers. They compared which bloggers behave more as 'starters' by computing the difference between the number of inlinks and outlinks of their blogs. Their experiments showed that it is possible to identify the top starters for a given query of several topic keywords in BlogScope. In this paper, we adopt the definition of the role of starter. In addition, we also propose connecters that link starters together as they are influential too.

Specially, Shetty and Adibi proposed the Entropy model to identify the most important nodes in a graph [9]. They dealt with the problem of finding leaders in a network. They built the graph so that nodes are representing persons or organizations and edges are representing actions they are involved in. They determined the important nodes by those who have the most effect of the graph entropy when they are removed from the graph. They used the event based entropy that has been similarly defined in [10]. Their experiment showed that comparing to conventional techniques such as betweenness centrality, this method leads to a better result. More important nodes can be discovered based on their effect on graph entropy in the ordered network. However, the graph entropy model claims its results on certain assumptions, like the evidence data is complete and with no noise.

Inspired by their ideas, we propose a method of measuring the influence of online posts through a refined graph entropy approach. In addition, the methods of Degree Measure and Shortest-path Cost Measure are exploited and integrated their results to identify the most influential posts. The details are discussed in Section 4. After the influential posts are identified, their authors are considered as potential influential users whose influence will be finally determined in the user graph model. In Section 6, we describe how to build the user graph model based on the post graph, meanwhile how to measure the users' influences from three aspects.

3   Graph Model of Online Posts

A lot of research work has been carried out in using graph methods to analyze the relationships between users on certain social networking websites [11, 12, 13, 14]. Here, we propose a general model of online posts which can be applied in different social networking sites while the user information is also taken into consideration.

A graph is defined as $G_v(V, E_v)$, where V is the set of posts and E_v is the set of directed edges which represent the relationship between those posts. Each post $v \in V$ can be described as a tuple of the form (n, t, u, c) where n is the node type, t is the timestamp, u is the author of the post and c is its content. Each directed edge $e \in E_v$ can be represented as $(v_i, v_j, p, w_{i,j})$ where v_i, v_j are nodes and e is an edge directed from v_i to v_j which means v_i is related to v_j, p specifies the type of relationship (either explicit or implicit), and $w_{i,j}$ is the weight of edge in range of $(0, 1]$ that measures the strength of their relationship. The relationship is directional and irreciprocal. It is defined that each post can only be related to (point to) earlier posts. Therefore, it is a directed acyclic graph and there should be of single edge connection between any two nodes as shown in Figure 1.

3.1   Types of Relationship

Explicit Relationship: It is given explicitly by the META information data collected from the social media platform, including the relationship of direct reply and some other forms (depending on the functions provided by the social network media, such as "share" on Facebook, "retweet" on Twitter and "citation" on forums). A relevance score $r_{i,j}$, which will be discussed later, is assigned to each edge from v_i to v_j, in order to calculate the edge weight. The score is set to 1 for all explicit relationships to represent full relevance. For example, $r_{i,j} = 1$ if v_i is a reply or retweet to v_j on Twitter.

Implicit Relationship: It is used to connect posts that are not directly related but talking on the same topic. The implicit relationship, $r_{i,j}$, indicates the degree of content relevancy from v_i to v_j that can be determined by measuring the content similarity score. The score should be in the range of $(0, 1]$. The conditions of building an implicit relationship can be different and depending on the features of the social networks applied on. In general, it is restricted by the time interval between two posts, as their relationship should weaken or dissolve when the time interval exceeds a certain time (called expiration time). For some forums in which only members within a group can see the posts of each other, the user's identity is also a restriction. For the blogging sites such as Facebook and Twitter, where one's posts can only be seen by friends or followers, the building of implicit relationships of posts is limited to their authors' friendship network.

3.2   Types of Posts

The type of a post is determined by the role it plays. The posts can be characterized in four types: root, follower, starter and connecter. Among them starters are certainly considered to be influential. Many researchers have tried to identify starters in a network as stated before. As for connecters, they are considered as bridges that connect two or more peaks in centrality analysis [15]. Also, a bridge node is also important in a network if it connects starters. Therefore, we define connecter to represent this type of nodes which are influential in a different way. Noted that in our definitions, follower, starter and connecter are referring to the type of posts.

Root: It is the first post discussing a topic or a subtopic within a certain period, so it is not related to any others. In the graph, roots are the nodes who are not pointing to others (with no out-degree).

Follower: It is a response (e.g. reply, comment or share) of a post or a new post talking on the same topic as another post before, which means it is explicitly or implicitly related to others. In a graph, followers are the nodes who are pointing to others (with some out-degree).

Starter: It is identified when it received a large number of explicit or implicit responses (followers); meanwhile the less it behaves as follower, the more it acts like a starter. In a graph, conversation starters are the nodes who point to a few but be pointed by many others, i.e., they are of high in-degree and low out-degree. Moreover, it is better for a starter to have followers also followed by many others. In a graph, it can be observed as having a high in-degree of followers.

Connecter: It connects two or more starters as a bridge, which means some starters will be disconnected without this node. It should be noticed that a post may play multi-roles at the same time. It is also possible that the roles of posts can change over time. The details of the identification of the node types will be discussed in Section 4.

3.3   Edge Weight

The weight assigned to each edge is the degree of relevance between two posts and high weight edges indicate strong relationships. Edge weight is measured by two factors: the content relevance and the time interval between posts.

$$w_{i,j} = \alpha_T \cdot r_{i,j} \tag{1}$$

An example is shown in Figure 1, where T is the time interval between the root post and its first reply. α_T is a factor used to diminish the relevance degree based on T. The details of calculating for content relevance between posts and edge weights have been introduced in our previous paper [22].

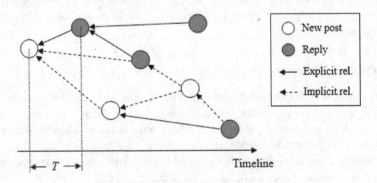

Fig. 1. A post graph with the timeline

4   Influence Measurements

4.1   Degree Measure

As mentioned above, the degree of a node can be used to identify starters. Since a starter is supposed to have a lot of followers and it is not a follower of many others, we first compute the difference between the in-degree and out-degree of each node. The in-degree of a node v denoted as $deg^+(v)$ is the sum of weight of the incoming edges incident to the node v, and the out-degree $deg^-(v)$ that is the sum of weight of its outgoing edges. The difference $d(v)$ is measured as one factor [8]:

$$d(v) = deg^+(v) - deg^-(v) \tag{2}$$

Another factor is the weighted average of its follower in-degrees to reflect the popularity of its followers:

$$s(v_i) = \frac{\sum_{v_j \in Fol(v_i)} w_{i,j} \cdot deg^+(v_j)}{\sum_{v_j \in Fol(v_i)} w_{i,j}}$$
$$= \frac{\sum_{v_j \in Fol(v_i)} w_{i,j} \cdot deg^+(v_j)}{deg^+(v_i)} \tag{3}$$

Then we can identify a node $v_i \in V$ as a starter when both $d(v_i)$ and $s(v_i)$ reach a threshold:

$$d(v_i) \geq \sigma_1 \wedge s(v_i) \geq \sigma_2$$

4.2   Shortest-Path Cost Measure

The basic idea of this method is to judge a node's influence by measuring how many other nodes would be affected and how much the influences are if the target node is removed from the graph. It should be noted that in a graph the relationship edges are built from later posts to earlier ones; conversely the influences traverse in reverse directions from earlier posts to later ones.

In our definition, a post should have influence on its followers, as the followers are responses (e.g. replies, citation and share) that are somehow activated by the original post (followee). These followers may also have influence on their own followers. As a result, a post may have indirect influences on its followers' followers, and so on. In a graph $G(V, E)$, the descendant set $Des(v)$ of a node $v \in V$ includes its followers directly pointing to it and other descendants that can reach it through paths. For every $v_d \in Des(v)$, there is at least one directed path from v_d to v in the graph.

If the path from node v_d to v_n is $(v_d, v_{d+1}, v_{d+2}, \ldots, v_n)$, the *relationship strength* from v_d to v_n can be measured as the accumulative weight:

$$W(v_d, v_n) = \prod_{i=d}^{n-1} w_{i,i+1} \qquad (4)$$

where v_i is pointing to v_{i+1} and $w_{i,i+1}$ is their edge weight. If more than one path from v_d to v_n exist, the maximum accumulative weight is taken as their relationship strength value. By doing this, the value of weight between any two nodes can be constrained in the range (0, 1]. The reason not to do summation and normalization of $w_{i,i+1}$ is that it will induce new weights with too small variance, which is difficult to differentiate afterwards. On the other hand, the ancestor set $Anc(v_d)$ of a node v_d is defined accordingly: $v_a \in Anc(v_d)$ when $v_d \in Des(v_a)$.

The algorithm of finding ancestors is similar to the one of finding the shortest path with respect to cost between nodes in a graph, except that we calculate the path cost as the product of the weights instead of the sum. It is assumed that each node would have influence on its descendants in the graph. To measure the influence of a node, we remove it from the graph and capture the change of path cost between these descendant nodes and their ancestors. The path cost $c(v_d)$ of a node v_d to its ancestors $v_a \in Anc(v_d)$ is the average of their relationship strength value:

$$c(v_d) = \frac{1}{|Anc(v_d)|} \sum_{v_a \in Anc(v_d)} W(v_d, v_a) \qquad (5)$$

Here we take the average in order to reduce the benefit for the nodes in later time, because later posts may have more ancestors. When a node v_i is removed from the graph, its adjacent edges are also removed. Its descendants $v_d \in Des(v_i)$ may be disconnected from some of their original ancestors. Even if they can reach their ancestors through other paths, their relationship strength may be weakened if the removed node is on their shortest path. Suppose v_i is on the path with the shortest cost between v_d and its ancestor $v_a \in Anc(v_d)$. After v_i is removed, a new path should be found with the new relationship strength value that $W'(v_d, v_a) \leq W(v_d, v_a)$. If no path can be traced between v_d to v_a, it means v_d is disconnected from v_a, and their relationship strength will be set to 0 ($W'(v_d, v_a) = 0$). If v_i is not on that path, the relationship strength between v_d and v_a will not change: $W'(v_d, v_a) = W(v_d, v_a)$.

Let $C(v_d, G, v_i)$ be the average shortest-path cost between the node v_d and its ancestors after removing v_i from the graph G. The influence of a node $v_i \in V$, $Inf_c(v_i)$, in the graph is then:

$$Inf_c(v_i) = \sum_{v_d \in Des(v_i)} (C(v_d, G, \varnothing) - C(v_d, G, v_i)) \qquad (6)$$

Compared to the degree measurement, this method considers multi-level relationship between posts, even if they are not on the same path. For example, as shown in Figure 2 (explicit relationship denoted by solid arrow and implicit relationship denoted by virtual arrow). Suppose node A is removed to see the influence on B and C. Then, B will be disconnected from any other nodes, while C can be still connected to D. Hence, A has a larger influence to B than to C. In this

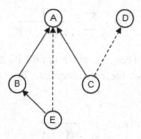

Fig. 2. An example graph of related posts

case, the influence measure of node A also considers the relationship between C and D, which is not considered in the degree measurement. Another advantage is the avoidance of duplicate counting on node E when measuring the influence of node A in multi-levels.

4.3   Graph Entropy Measure

Based on the graph model proposed, a graph can be considered as an ordered network with the node types of root, follower, starter and connecter defined. Shetty and Adibi [9] showed their success in finding important nodes through graph entropy in an ordered network. The graph entropy can be defined differently for various problems and we adopted a similar approach as in Dehme [18]. In their work, the entropy of a network is defined by using the local information graph, where metrical graph properties are used for defining information functional of each vertex.

Consider a graph with arbitrary node labels. In order to determine the probability value for each node so that it can be used to calculate the graph entropy, we first need to define the local vertex functional. Generally, the information functional is used to quantify structural information based on a given probability distribution. In our case, we define the information functional as the centrality of nodes.

For the graph $G = (V, E)$ where $v_i \in V$, graph entropy is defined by:

$$E(G, P) = \sum_{i=1}^{|V|} p(v_i) \log(1 / p(v_i))$$ (7)

The probability for each node is defined as:

$$p(v_i) = \frac{f(v_i)}{\sum_{j=1}^{|V|} f(v_j)}$$ (8)

f represents an arbitrary information functional. Unlike traditional centrality measurement, such as closeness centrality, betweenness centrality and eigenvector centrality, in our model the centrality of a node only looks at the nodes that point to it or can be reached through paths. Recalling the term of "descendant" that is defined in last section, a node's descendants is used to measure its distance-weighted centrality.

$$f(v_i) = \sum_{v_d \in Des(v_i)} \frac{1}{d(v_d, v_i)} \qquad (9)$$

$d(v_d, v_i)$ is the distance between the node v_i and its descendant v_d. If there is an edge that directly links to them, their distance can be calculated as the reciprocal of the edge weight.

$$d(v_d, v_i) = \frac{1}{w(v_d, v_i)} \qquad (10)$$

Otherwise, if v_d can reach v_i through a path ($v_d, v_{d+1}, v_{d+2}, \ldots, v_i$), then the distance between v_d to v_i will be the sum of edge distance along the path. In case that more than one path exists, the shortest path distance will be taken.

The steps of measuring node influence through graph entropy are shown below.

1. Compute the entropy of each node v_i as:

$$E(i) = -p(v_i) \cdot \log(p(v_i)) \qquad (11)$$

2. Remove v_i and its edges from the graph
3. Calculate the entropy of remaining graph as $EN(i)$
4. Calculate the influence of node v_i as:

$$Inf_e(v_i) = \frac{EN(i)}{\log(EN(i) / E(i))} \qquad (12)$$

The formula (12) is referred from [9], which proved to be able to identify important nodes in the network built of Enron (company) emails. We adopt it to measure the influence of node v_i by $E(i)$ and $EN(i)$, and try to find nodes which have higher centrality and more effect in the graph after they are removed from the graph.

4.4   Identify Influential Posts

To find the influential nodes, we ranked the nodes based on their influence scores from different measurements. Starters and connecters can be identified first as the preliminary result. Starters are determined by degree measure, and connecters are identified by using the other two methods. In our proposal, a connecter should fulfill two conditions: (i) Have a higher rank in the measurements of shortest-path cost or graph entropy. (ii) Connect two starters by different authors.

As we have defined influence from the aspects of starter and connecter, the influential nodes are either starters or connecters. Based on the combination of three measures, we are able to determine the most influential posts. The following are the heuristic used to determine the influential posts and potential influential users:

1. Remove the starters from the list of influential nodes if they are ranked low in all measurements.
2. Remove the connecters from the list accordingly if their connected starters are not influential.

3. Consider the connecters not critically influential if there are candidate connecters between the same set of starters.
4. Other starters and connecters are considered as influential posts, and their authors are considered as potential influential users.

5   User Graph Model

Although the influential starters and connecters are identified from the post graph, we still have the problem on determining the influential users. Consider the cases that (i) for a starter many of its followers are actually from a small group of users (one user can reply several times); (ii) a connecter links with two starters who have a large set of common follower users. In these cases the influence may be wrongly judged in the post graph model. Therefore we proposed the user graph model to refine the influence measures of potential influential users.

A user graph can be converted from the post graph. However, we are not going to build a complete user graph due to high computational complexity. As we are more interested in users who have made influential posts, we select the authors of starters and connecters in the post graph as seeds, then look at their neighbours and finally find possible connections between distant starters. The process of converting post graph to user graph will be discussed in next session.

The user graph is defined as $G_u(U, E_u)$, where U is the set of users, E_u is the set of directed edges which represent the relationship between users. Each node $u_k \in U$ is the author of post v_i^k in the Post Graph G_v.

Node types: There are three types defined in the user graph: starter, connecter and follower. Each node can belong to one or more types. At first, the type of a user is the summation of types of his posts. For example, starter users are the authors of starter posts identified in the post graph. But connecter is a special type. The author of a connecter in the post graph may no longer play the same role in the user graph. On the contrary, some new nodes could be detected as connecters in the user graph, even though none of their posts connect two starters in the post graph. Therefore the type of connecter will be determined after the graph conversion and measurement.

Edge types: $e(u_k, u_j) \in E_u$ is the edge directed from u_k to u_j representing that u_k is related to u_j, which means u_k has replied or responded to u_j either explicitly or implicitly (as defined in the post graph model). Besides, there are another type of virtual edge $e'(s_k, s_j)$ defined between two starters, to represent the directed path from s_k to s_j. The virtual edges are built when there are at least one directed paths between two starters, and their distance is very long. In this case, we will keep the shortest path length as the weight of virtual edge. The nodes on the paths are not important so it is not necessary to show them in the user graph. Moreover, the edges are directed and will be considered as two separate edges when they link two nodes in opposite directions.

Edge weight: $w(u_k, u_j)$ is the weight of edge $e(u_k, u_j)$ that measures the strength of their relationship. It is affected by the times of interactions and the relevancy of their conversations. For the virtual edges defined to link two starters, the edge weight $w'(s_k, s_j)$ is calculated as the shortest path length from s_k to s_j in user graph as described before.

6   Graph Conversion and Measures

In order to capture the influence of users, a user graph is needed. The next step is to convert the post graph to the user graph. Instead of processing the complete graph, we use a biased sampling method starting with potential influential users, who have posts as starters or connecters identified in the post graph. Then we propose several measurements to capture the influences of u-starters and u-connecters in different respects. (The terms "**u-starter**" and "**u-connecter**" are used to refer to the starters and connecters in user graph model.) For the rest of the section, we discuss the details of how to convert the post graph to user graph and measure the influences of u-starters and u-connecters. Figure 3 shows the overall flow of the operations and measurements on user graph.

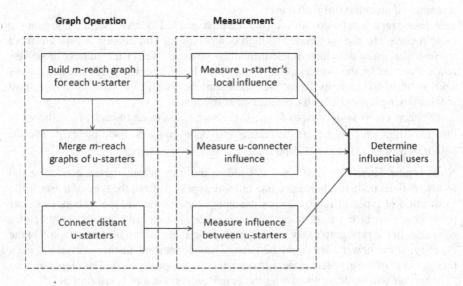

Fig. 3. Workflow of graph operation and measurement

6.1   Build *m*-Reach Graph for Each u-Starter

In the post graph model, a starter is observed when it has obtained a large amount of followers and descendants. However, it is hard to measure its influence on users, because one user may write a number of posts, or reply several times within a discussion. Moreover, if a user has several posts as starters, it is necessary to consolidate all the followers and descendants in terms of users. For this reason, we need the conversion from the post graph to a user graph where each user is represented as one node. But if the user graph is directly built for all discussions from different u-starters, some of their descendants will be merged and their influence may not be accurately judged. Therefore we first built an m-reach user graph for each u-starter in order to capture its local influence.

M-Reach Graph

"M-reach" is a measure defined by Borgatti[19] that counts the number of unique nodes reached by a given node in m links or less. In our user graph, $g^m(u_k)$ is u_k's m-reach graph which consists of nodes that can reach u_k via a path of length m or less. Here the path length is defined as the number of hops to go though without consideration of edge weights.

Discussion Thread and Discussion Chain

In the post graph, a starter together with its descendants forms a discussion thread. In the Post-reply Opinion Graph by Memon and Alhajj [20], they clearly defined the discussion chain which is different from discussion thread: "The discussion chains consist of the paths in the graph whose starting node is a root and ending node is a leaf when we inverse the direction of the edges." In Figure 4, a post-reply graph shows the difference between discussion chains and threads.

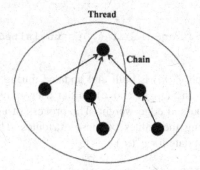

Fig. 4. Discussion threads and discussion chains

Algorithm of Building m-Reach Graph

Suppose the set of starters found in post graph $G_v(V, E)$ is $S \subset V$, each starter (post) $s_i^k \in S$ has an author u_k, then u_k is a u-starter. An m-reach user graph $g^m(u_k)$ will be built for each u-starter u_k. For each starter s_i^k by user u_k, the discussion thread in post graph will be converted to user graph $g^m(u_k)$. Here the value of m will be determined during the experiment.

In order to keep the information of distances (as defined in Section 4.3) from the starter to its descendants in a discussion chain, depth-first search (DFS) starting from s_i^k is conducted in the post graph G_v. For each descendant v_a^x of s_i^k (with authors u_x and u_k respectively), the shortest distance between v_a^x and s_i^k is notated as $d_a^{(x, k)}$.

While the distance considers edge weights in the post graph, the path length is defined differently for "m-reach". That is, the path length from v_a^x to s_i^k is the minimum number of distinct users on the path for v_a^x to reach the starter s_i^k. It is represented as $m_a^{(x, k)}$, and used to control the depth of searching. Suppose the value of m is given as m_0, the pseudo code is shown below:

```
1  for each starter sᵢᵏ by user uₖ
2      label sᵢᵏ as visited, set mᵢ^(k, k) to 0
3      let S be a stack
4      S.push(sᵢᵏ)
5      while S is not empty
6          vₐˣ := S.top()
7          for each vₐˣ's unvisited follower v_b^y in G_v
8              label v_b^y as visited
9              if there is a visited node v_o^y with author u_y
10                 m_b^(y, k) := m_o^(y, k)
11             else
12                 m_b^(y, k) := mₐ^(x, k) + 1
13             if m_b^(y, k) <= m₀
14                 update g^m(uₖ) with node v_b^y and edge e(v_b^y, vₐˣ)
15                 S.push(v_b^y)
16             continue at 5
               /*Reset the node vₐˣ as unvisited after all its followers
                 are visited, so that it can be visited in other path*/
17         delete mₐ^(x, k) and label vₐˣ as unvisited
18         S.pop()
```

The m-reach user graph $g^m(u_k)$ is built and updated during the process of DFS in the post graph (as shown in Step 14 above). In our user graph model, there are two basic attributes: node type and edge weight. The process of updating m-reach graph actually refers to changing the values of these attributes. The pseudo code below shows how to build and update for $g^m(u_k)$.

```
1  add node uₖ with type (starter) in g^m(uₖ)
2  for each node v_b^y and edge e(v_b^y, vₐˣ) obtained from DFS in G_v
3      if v_b^y is not visited
4          if there is no user node u_y in g^m(uₖ)
5              add a new node u_y in g^m(uₖ)
6          add v_b^y's type in u_y's type
7      if e(v_b^y, vₐˣ) is not visited
8          if there is no edge from u_y to u_x in g^m(uₖ)
9              build the edge e(u_y, u_x)
10             w(u_y, u_x) := w(v_b^y, vₐˣ) /*initialize edge weight*/
11         if there is an edge e(u_y, u_x) in g^m(uₖ)
12             w(u_y, u_x) := w(u_y, u_x) + w(v_b^y, vₐˣ) /*update edge weight*/
```

An example of building m-reach graph is illustrated in Figure 5: (a) is a post graph showing the relationship between six posts (node 1 to 6) with four authors A, B, C and D; (b) is the m-reach user graph converted from (a), each node represents a user with its post IDs labeled in the bracket. In (a), node 1 is a starter with author A, the other nodes are its descendants. The edge weights are labeled beside the edges.

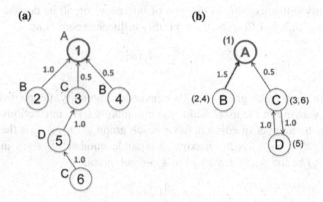

Fig. 5. Example of building m-reach user graph from post graph

Suppose we want to convert this post graph to an m-reach user graph with $m = 2$, node 2 and 4 are in 1-reach, and they belong to the same user B, so the two nodes are merged into one node B in (b), while the weight of the edge B→A is the sum of the weights for 2→1 and 4→1. If we look at the chain of nodes 1, 3, 5 and 6, the path lengths for the descendants to reach the starter are: $m_3^{(C, A)} = 1$; $m_5^{(D, A)} = 2$; $m_6^{(C, A)} = 1$. It should be noticed that the author of node 6 is C, which is the same as node 3, so the path length for node 6 is reduced to 1. If node 6 has followers by other users, those followers will have the path length equal to 2, therefore they will also be considered within m-reach.

As for the connecters, because they are defined as bridges to link with starters, they are certainly in 1-reach to a starter. This means all the connecters will be included some starter's m-reach graphs as long as $m \geq 1$.

6.2   Measure the Local Influence of u-Starter

The m-reach graph can be used to measure the local influence of u-starters. We proposed three measures to calculate a u-starter's influence in its m-reach graph from three aspects.

The distance-weighted centrality of a node has been defined in (10). It is a measurement that counts the number of its descendants with the weight reciprocal to their distances. The distance information can be obtained from the post graph and used for calculating the influence score of a u-starter on its descendants. It is defined that for the u-starter u_k, the maximum value of its influence on each user u_x is 1. $d_a^{(x, k)}$ is the shortest distance between v_a^x and s_i^k in G_v, then the influence score of u_k on u_x is:

$$I_k(u_x) = \text{Min}\left(\sum_{v_a^x \in Des(s_i^k)} \frac{1}{d_a^{(x,k)}}, 1 \right)$$ (13)

The centrality influence of u_k is the sum of influences on all its descendants in the m-reach graph $g^m(u_k)$. Let $C(u_k)$ be the centrality influence score of u_k.

$$C(u_k) = \sum_{u_x \in g^m(u_k)} I_k(u_x) \tag{14}$$

The users in an m-reach graph actually consist a community. Graph density is used to measure how many of the users within the community have interactions with many others. $|E_k^m|$ is the number of edges in the m-reach graph $g^m(u_k)$, $|V_k^m|$ is the number of nodes, and $|V_k^m|(|V_k^m| - 1)$ is the maximum possible number of edges in a directed graph. Let $D(u_k)$ be the graph density of u_k's m-reach graph.

$$D(u_k) = \frac{|E_k^m|}{|V_k^m|(|V_k^m| - 1)} \tag{15}$$

The third factor considers how strong the interactions are in the u-starter's community. It is measured by the sum of weights of all the edges in the m-reach graph.

$$N(u_k) = \sum_{g^m(u_k)} w(u_y, u_x) \tag{16}$$

The three factors are summarized into an influence score of the u-starter u_k using the formula below:

$$M_S(u_k) = \frac{\alpha \cdot C(u_k) + \beta \cdot D(u_k) \cdot (|V_k^m| - 1) + \gamma \cdot N(u_k) / |V_k^m|}{\alpha + \beta + \gamma} \tag{17}$$

Each factor is weighted depending on user's need and the feature of real data. Besides, there are normalization factors associated with $D(u_k)$ and $N(u_k)$.

6.3   Merge m-Reach Graphs of u-Starters

Since the m-reach graph is built for each u-starter separately, it is possible that one user exists in several m-reach graphs. In this step we would like to merge the common user nodes as well as their edges in different m-reach graphs. Figure 6 shows an example of merging two u-starters' m-reach graphs ($m = 2$).

The process of merging nodes includes the combination of node types for the same user. Their associative edges will be added together. The edge weight will remain the same if there is only one edge from one user to another. In the cases that more than one edges exist between two users with the same direction, these edges will be merged and the maximum edge weight among them will be taken as the merged edge weight. They are associative operations. So the overall action of merging is associative, which means the result is unique no matter what the merging sequence is.

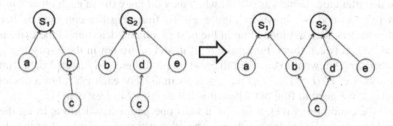

Fig. 6. Merge 2-reach graph of two u-starters

6.4   Measure the Influence of u-Connecter

After merging the m-reach graphs of different u-starters, the u-connecters should be in a graph joining all m-reach graphs from u-starters they connect.

First, if a u-connecter links two u-starters that already directly connected in the user graph, it is determined no longer a connecter as there is no need to have a connecter here.

For the existing u-connecters, there should be a way to measure and compare their influences. We would like to adopt the method of Shortest-path Cost Measurement used in the post graph model which approved useful to identify connecters. The basic idea is to remove the u-connecter from the user graph and measure the impact on the influence propagation from the u-starters. The same formula is used to calculate the influence of a u-connecter u_k:

$$M_C(u_k) = \sum_{u_d \in Des(u_k)} (C(u_d, G_u, \varnothing) - C(u_d, G_u, u_k)) \tag{18}$$

However, this formula has a different meaning, as the ancestors are replaced with a u-starter here. Let $C(u_d, G_v, u_k)$ be the sum of the *relationship strength* (as defined in Section 4.2) from u_d to the u-starter after removing u_k from the graph G_v. This u-starter should be the parent of the u-connecter u_k. In the case that u_k has several u-starters as parents, the sum of measuring results for several u-starters will be taken as the final influence score of u-connecter u_k.

Besides the existing ones, some new u-connecters may be found as a broker to link two u-starters (one is his parent and the other is his child in the user graph). We can also use the above method to measure their influences. But the new u-connecters do not have a post as connecter in the post graph, which means they have not behaved as connecter within a discussion, they are only considered as potential connecters who should have the ability but have not conducted.

6.5   Connect Distant u-Starters

After merging the m-reach graphs, still there may be disconnected subgraphs, or some isolated m-reach graphs of u-starters. In order to connect them and discover inter-starter influences, we built virtual edges between distant u-starters.

The u-starters are defined as distant when they do not exist in each other's m-reach graphs (e.g. S_1 and S_3 shown in Figure 7). To find possible connections between distant u-starters, we first looked up in the post graph and determined the existence of directed path between them. For example, if u_j and u_k are not in the m-reach graph of each other, first we want to check in the post graph if there is a directed path from u_j's post to u_k's. Let v_i^k ($i = 1,2,3,...$) be u_k's posts in G_v. For each v_i^k, it has a descendant set $Des(v_i^k)$. We need to find out whether u_j has a post v_d^j in $Des(v_i^k)$.

Once the condition is met, it means at least one path exists from u_j to u_k, then we will build an virtual edge from u_j to u_k. The edge weight is calculated as the shortest path length (number of distinct users) between them. Similarly, we can check if the inverse path from u_k to u_j exists. The edges are considered as different in opposite directions.

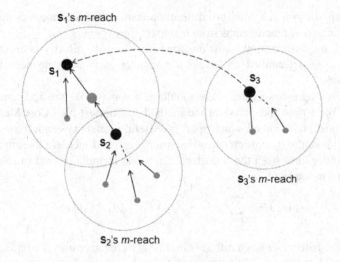

Fig. 7. Connect distant u-starters (from S_3 to S_1)

6.6   Measure the Influence between u-Starters

After all the possible virtual edges are built, the influence of one u-starter s_k on another one s_j can be calculated by:

$$I_k'(s_j) = \begin{cases} 1 & \text{if } s_j \text{ is in } g^m(s_k) \\ \min(\dfrac{m}{w'(s_j, s_k)}, 1) & \text{if } e'(s_j, s_k) \text{ exists} \end{cases} \qquad (19)$$

where the value of m will be determined in the experiment. An example is shown in Figure 7 that S_1, S_2 and S_3 are u-starters, and S_1 has influence on S_2 and S_3. The influence of S_1 on S_2 counts as 1 as S_2 is in S_1's m-reach graph, while S1's influence on S3 is measured by the second formula. Finally, the influence of the u-starter s_k on other starters is the summation of influences on each one:

$$M_I(s_k) = \sum I_k'(s_j) \qquad (20)$$

7   Experiment

7.1   Case Study for Post Graph Model

Our proposed model can be applied for different social media. Both explicit and implicit relationships can be identified between text-based posts. We chose Twitter to conduct the experiments as it has many users and its data are easy to collect.

In order to find the most influential posts and their respective authors during the information diffusion within a topic, we select a general user (neither famous people nor public media) who has written some posts on a topic, find the user's friends who have responded to the posts or also talked on this topic, then dig out the friends of friends and so on. General sampling method is not suitable here, because we need the data from users with more connections between them so that the graph can be well formed. Tweet data are collected on the topic of "Steven Jobs and iPhone 4s". The keyword set is defined as {"iPhone 4", "iPhone 4s", "iPhone 5", "iPhone Mini", "Steve Jobs", "Apple", "ios 5", "Siri"}. Table 1 gives the data description for the experiment.

Table 1. Description of data

Platform	Twitter
Topic	Steven Jobs and iPhone 4s
Time	11/10/2011 - 31/10/2011
Location	Hong Kong
No. of users	158
No. of tweets	211

Preliminary Results

Preliminarily, starters and connecters can be found after the three influence measurement methods are applied. As mentioned before, degree measure can be used to identify starters. Two factors are calculated: (i) the degree of each node $d(v)$; (ii) the weighted average of its follower in-degrees $s(v)$. The top nodes that $d(v) + s(v) > 2$ are selected. The results are plotted in the diagram shown in Figure 8.

It is observed that the results of the two factors are not aligned most of the time. The reason is that a node with higher degree should have more followers, and it becomes difficult for all its followers to have a high in-degree. On the contrary, there exist some nodes with only a few followers, but most of the followers have high in-degree. These nodes can be detected by high score of $s(v)$. For our work, we finally selected the 10 nodes with $d(v) > 3$ and $s(v) > 0.1$ as starters (Node 1 – 10 labeled in Figure 9).

As for the connecters, we integrate the results from Shortest-path Cost Measure (SCM) and Graph Entropy Measure (GEM). After calculating the influential scores $Inf_c(v_i)$ and $Inf_e(v_i)$, all the nodes are ranked. After examining the top ranking nodes, besides the found starters, other nodes which connect starters are considered as connecters (Node 11 – 20 in Figure 9). The connecters discovered by each method are listed below in ranking order.

— SCM: 11, 14, 13, 12, 15, 16, 20, 17 (nodes labeled in Figure 9)
— GEM: 11, 14, 12, 15, 13, 16, 20, 17, 18, 19 (labeled in Figure 9)

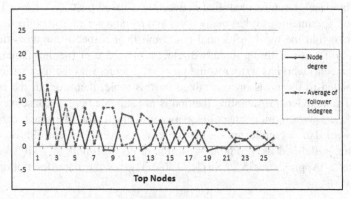

Fig. 8. Degree measures

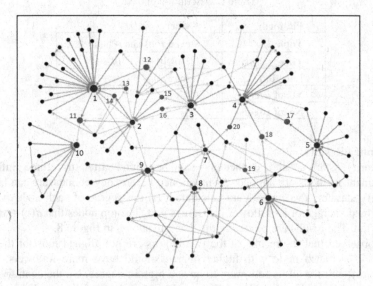

Fig. 9. Graph of starters and connecters

Discussion and Final Results

In comparison, SCM only identifies 4 starters in its top 10 ranking nodes, and is able to find all starters in top 21; while GEM can find 7 starters in top 10 and all starters in top 14. It is because GEM looks into both node entropy $E(i)$ and remnant graph entropy $EN(i)$ in calculating the influence score, which is aimed to achieve high node centrality as well as large effect in the graph after removal. As for the SCM algorithm, we can see that its influence score is in the range from 0 to the number of the node's descendants. There is no difference between its close followers and distant descendants when measuring a node if the weights are all 1. As a result, it is more likely to find the nodes with more descendants, whereas GEM can find the nodes with more ancestors or descendants.

Finally, we can find the most influential posts considering the results of all measurement. For the starters, node 7 and node 8 by different authors are ranked low by SCM and GEM, so they are not considered to be influential in the final result. Since node 7 is not influential any more, we look at the connecters 19 and 20 that connect node 7. It is found that they also have relatively low rankings. Therefore they are also removed from the influential list.

Noted that not every node that connects two starters can be a connecter, the connecters are detected by the two measurements, which means their removal from the graph will have a certain impact on the information transmission, and they should have some followers to make them more influential. In Figure 9, we can see that nodes 13 and 14 are actually connecting the same starters 1 and 2, and so are the nodes 15 and 16 which connect starters 2 and 3. In this case, we consider them not to have critical influences.

7.2   Case Study for User Graph Model

In order to compare the results in finding influential users in post graph and user graph, a larger data set is needed for experiment. In this case study, we collected more than 1700 tweets from 915 users, on the topic of "Sichuan Lushan earthquake" (an earthquake happened in China on April 20, 2013) and "H7N9 influenza". More description on the data set is shown in Table 2.

Table 2. Description of data

Platform	Twitter
Topic	"Sichuan Lushan earthquake", "H7N9 influenza"
Time	31/03/2013 - 30/04/2013
Location	China, Hong Kong, Japan

Table 3. Top 5 influential users in post graph

Rank	1	2	3	4	5
No. of starters and connecter	6	6	5	3	2

Influential Users in Post Graph

Similar actions are taken as in previous case study to identify starters and connecters in the post graph. Finally, 49 starters and 5 connecters are found in the posts. Some starters or connecters are actually written by the same authors, so we identified 29 users as influential in total. If we rank the influential users found in post graph according to the number of starters/connecters they have, the top 5 users are listed below: (For those with the same number of starters and connecters, they are ranked based on the highest ranking of their posts.)

In order to justify our user graph model, we converted the post graph into user graph, and then measured the users' influences in the user graph of two types: u-starter and u-connecter.

U-starter Influence

The local influence of a u-starter is measured by three factors as shown in (18). In this experiment, we put more weight on the centrality measure, set $\alpha = 2$, $\beta = 1$, $\gamma = 1$. The value of m is decided by the distance between close starters in the post graph. We tried to make more m-reach graphs contain only one starter, meanwhile have common descendant nodes so that they are connected after the merging operation. For the cases that the starters are far away from each other, we suggested the m value not larger than 5. The influence between u-starters is also taken into consideration. When some u-starters have similar local influence, their ranking will be judged by the inter influence measure. After all, the ranking of influential starters is a little different from that in post graph. In top 5 influential users found:

- Top 2 users keep the same.
- A new influential user is identified on rank 3 in user graph.
- The user on rank 4 in post graph does not rank on top in user graph.

The new influential user found in the user graph only has one post as starter. But this starter has a large number of followers, and these followers have interactions with each other, which makes its local influence score higher. Figure 10(a) and 10(b) shows the post graph and corresponding user graph of this node and its descendants within 5-reach. For the user falling off the top 5 list, the main reason is that his followers or friends are from a small community, and there are no connecters to propagate their discussion to another community.

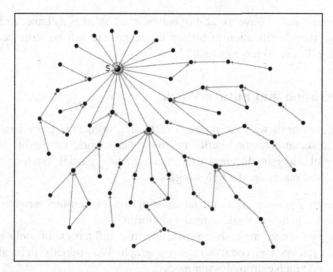

Fig. 10(a). Post graph of an influential starter

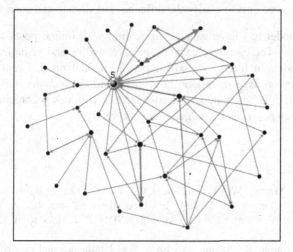

Fig. 10(b). User graph of the starter

U-Connecter Influence

In the post graph there are 5 connecters identified. But after the graph is converted into user graph, it is found that 2 of them are not connecters anymore, because the u-starters they link with are directly connected. The remaining 3 u-connecters are determined to be influential users.

However, 1 new u-connecter is found in the user graph, who link with 2 different u-starters. As stated above, it is only considered as a potential connecter. The result proves that in the post graph the connecters already identified can be refined and some new connecters may be found. The new connecters are not that influential as they are just supposed to have the ability but have not acted as a connecter in our data set. Therefore the identification of influential connecters will be more accurate and complete if the data set is large enough.

8   Conclusion and Future Work

In this paper, we dealt with the problem of finding influential users based on their interactions in social networks. Different from other's work, we tried to identify the most influential users in different roles through their posting on the same topic. Additional contributions are the following:

- We proposed a general graph model showing the relationship between posts that can be applied in different social media platforms.
- We presented three methods to measure the influences of online posts to distinguish starters and connecters in the graph. We specially defined the node centrality and graph entropy for our model.
- We converted the post graph to user graph using biased sampling, and proposed different measurements to clarify the influences of starters and connecters.

Our graph model has its advantage in dealing with online posts and users with more interactions. Therefore we carried out case studies and visualize the graph to validate the model. In future, we will apply the model on different social media platform to carry on experiments on larger data set. Furthermore, this model can be more effective if it is integrated with advanced text mining techniques, so that the relevance between posts can be judged more accurately.

References

1. Bakshy, E., Mason, W.A., Hofman, J.M., Watts, D.J.: Everyone's an Influencer: Quantifying Influence on Twitter. In: Proceedings of the 4th ACM International Conference on Web Search and Data Mining, WSDM 2011, Hong Kong, China, pp. 65–74 (2011)
2. Bakshy, E., Karrer, B., Adamic, A.: Lada. Social Influence and the Diffusion of User-Created Content. In: 10th ACM Conference on Electronic Commerce, Stanford, California. Association of Computing Machinery (2009)
3. Leavitt, A., Burchard, E., Fisher, D., Gilbert, S.: The Influentials: New Approaches for Analyzing Influence on Twitter. Web Ecology Project (2009), http://tinyurl.com/lzjlzq
4. Klout Score, http://klout.com/home
5. Twinfluence, http://twitterfacts.blogspot.com/2008/10/twinfluence.html
6. Weng, J., Lim, E., Jiang, J., He, Q.: TwitterRank: Finding Topic-sensitive Influential Twitterers. In: Proceedings of the 3rd ACM International Conference on Web Search and Data Mining, WSDM 2010, pp. 261–270 (2010)

7. Hansen, D.L., Shneiderman, B., Smith, M.A.: Visualizing Threaded Conversation Networks: Mining Message Boards and Email Lists for Actionable Insights. In: An, A., Lingras, P., Petty, S., Huang, R. (eds.) AMT 2010. LNCS, vol. 6335, pp. 47–62. Springer, Heidelberg (2010)

8. Mathioudakis, M., Koudas, N.: Efficient Identification of Starters and Followers in Social Media. In: Proceedings of the 12th International Conference on Extending Database Technology: Advances in Database Technology, EDBT 2009, pp. 708–719 (2009)

9. Shetty, J., Adibi, J.: Discovering Important Nodes through Graph Entropy. In: The 11th ACM SIGKDD International Conference on Knowledge Discovery and Data Mining (2005)

10. Nobel, C., Cook, D.J.: Graph-based anomaly detection. In: Proceedings of the 9th ACM SIGKDD International Conference on Knowledge Discovery and Data Mining, pp. 631–636 (2003)

11. Ilyas, M.U., Radha, H.: A KLT-inspired Node Centrality for Identifying Influential Neighborhoods in Graphs. In: Conference on Information Sciences and Systems, pp. 1–7 (2010)

12. Tang, L., Liu, H.: Graph Mining Applications to Social Network Analysis. Managing and Mining Graph Data. In: Managing and Mining Graph Data, pp. 487–513 (2010)

13. Sala, A., Cao, L., Wilson, C., Zablit, R., Zheng, H., Zhao, B.Y.: Measurement-calibrated Graph Models for Social Network Experiments. In: Proceedings of the 19th International Conference on World Wide Web, WWW 2010, pp. 861–870 (2010)

14. Wilson, C., Boe, B., Sala, A., Puttaswamy, K.P.N., Zhao, B.Y.: User interactions in social networks and their implications. In: Proceedings of EuroSys, pp. 205–218 (April 2009)

15. Scott, J.: Centrality and Centralization. In: Social Network Analysis: a Handbook. SAGE Publications, London (2000)

16. Lipkus, A.H.: A proof of the triangle inequality for the Tanimoto distance. J. Math. Chem. 26(1-3), 263–265 (1999)

17. Sun, B., Ng, V.T.Y.: Lifespan and Popularity Measurement of Online Content on Social Networks. In: Social Computing Workshop of IEEE ISI Conference, pp. 379–383 (2011)

18. Dehme, M.: Information processing in complex networks: Graph entropy and information functionals. In: Applied Mathematics and Computation, vol. 201, pp. 82–94 (2008)

19. Borgatti, S.P.: Identifying Sets of Key Players in a Social Network. Comput. Math. Organiz. Theor. 12, 21–34 (2006)

20. Memon, N., Alhajj, R.: From Sociology to Computing in Social Networks

21. Tunkelang, D.: A Twitter Analog to PageRank (2009),
http://thenoisychannel.com/2009/01/13/
a-twitter-analog-to-pagerank/

22. Sun, B., Ng, V.T.Y.: Identifying Influential Users by Their Postings in Social Networks. In: Proceedings of the 3rd International Workshop on Modeling Social Media, pp. 1–8 (2012)

Modeling a Web Forum *Ecosystem* into an Enriched Social Graph

Tarique Anwar[1] and Muhammad Abulaish[2]

[1] Centre for Computing and Engineering Software Systems
Swinburne University of Technology, Melbourne, VIC 3122, Australia
tanwar@swin.edu.au
[2] Department of Computer Science
Jamia Millia Islamia (A Central University), New Delhi 25, India
mAbulaish@jmi.ac.in

Abstract. This paper considers the community interactions in online social media (OSM) as an *OSM ecosystem* and addresses the problem of modeling a Web forum ecosystem into a social graph. We propose a text mining method to model cross-thread interactions and interests of users in a Web forum ecosystem to generate an enriched social graph. In addition to modeling *reply-to* relationships between users, the proposed method models message-similarity relationship to keep track of all similar posts resulting out of deviated discussions in different threads. Although, the proposed graph-generation method considers a *reply-to* relation as the primary means of linkage, it establishes links between clusters of similar posts instead of links between individual users, and the linkages between users can be derived from the existing linkages between clusters. The method starts with linking posts in each thread individually by identifying *reply-to* relationships, and applies an agglomerative clustering algorithm based on similarity of posts across the forum to group all posts into different clusters. Finally, relations between each pair of individual posts are mapped to create a link between clusters containing the posts. As a result, the generated social graph resembles a network of clusters that can also be presented at the granule of users who authored the posts to generate a social network of forum users, and at the same time it keeps information for all other users with similar interests.

Keywords: Social media mining, Web forum ecosystem, Social graph generation, Agglomerative clustering.

1 Introduction

Since the inception of Web 2.0, it is increasingly getting crowded with Web users and explosive contents generated by them at a tremendous rate, which characterizes the Web as extremely dynamic and diverse in nature [6]. Nowadays, Web 2.0 applications are endorsing a paradigm shift in the way contents are generated on the Web [29], where users are getting space to generate contents by themselves. A significant percentage of Web users are frequently participating

M. Atzmueller et al. (Eds.): MUSE/MSM 2012, LNAI 8329, pp. 152–172, 2013.

in various ways to generate Web contents [21]; social networking sites (SNS) such as facebook, twitter, myspace, etc., being the most common of them are intruding rapidly into our lives [24]. Thus, it conduces us to categorize Web contents into the *proprietary contents*, adhering to some well-defined structure, and the *user-generated contents*, which are highly unstructured and noisy [11]. The group of Web-based applications (a division of cyberspace) that build on the ideological and technological foundations of Web 2.0, and allow the creation and exchange of user-generated contents is said to be *online social media* (OSM) [23].

1.1 The OSM *Ecosystem*

Even though the term contains the word "media", it has very little to do with the traditional information media. Rather, it is more inclined towards the other word "social" (derived from the ties of social relationships) and provides a mechanism for the audience to socialize themselves by mingling with others [9]. OSM is evolving as a powerful tool for people to connect, communicate and interact globally on topics of common interest which take place in various forms ranging from complicated and obscured ones to simple and conspicuous ones. Some instances of these interactions are, i) posting a comment on a facebook update, ii) liking a link shared on a friend's facebook wall, iii) following someone or being followed by someone on twitter, iv) commenting or replying a blog post, v) participating in a discussion thread on a Web forum, vi) liking or disliking a youtube video, and so on. Even a layman can easily notice the conspicuous relationships in the above instances, but other obscured relationships in them can hardly be noticed even by an expert analyst. For example, let us assume three social media users, u_1, u_2 and u_3. Suppose u_1 comments on a thread initiated by u_2, and u_3 is the user to whom u_2 replies every time she asks any question in a thread. In this scenario, the relationships between u_1 and u_2 as well as u_2 and u_3 are comparatively much more noticeable than the one between u_1 and u_3. Usually the relationship that an analyst tries to accentuate depends on the type of interaction being focused, where interaction type can be any of the above mentioned or similar instances. These kinds of relationships established among users on the Web create a healthy, social and collaborative *ecosystem* for various community practices. In [21], Jones and Fox observed that nowadays people use the Internet more often to socialize themselves through social media than other activities. For example, when a person without adequate technical knowledge about cars, finds some fault in the gearbox of his car, he initiates a thread on a forum explaining its unexpected symptoms and asks for helpful suggestions. Very soon the thread gets multiple replies by unknown users who share their own experiences with a similar problem and suggest probable solutions that could be helpful to him. In one perspective, the replies in the thread are nothing more than an assistance to the thread initiator, which is clearly visible to all. Can the series of replies, personal experiences and suggestions also be meaningful for any other reason and/or other person? A ponder on this question brings about a considerable number of other perspectives beyond the only apparent one, that makes us realize the importance of such interactions. The general objectives in this area span over characterizing user behaviors and interactions, analyzing established social relationships among them, and extracting information from the text contents of actual discussions. However, Choudhury *et al.* [10] pointed out that the excitement of information extraction researchers resulting from the overwhelming and explosive growth in user generated Web contents is overshadowing two authoritative and related problems in this area. First one is the *inference problem*, which states that the real social relationships or ties always remain obscured and therefore must be deduced from the unobscured events to bring it into notice. In the example mentioned above, the obscured tie between u_1 and u_3 is to be deduced from their unobscured interactions with u_2. The other problem pointed out is the *relevance problem* according to which a social network actually is a blend of multiple social networks, each one based on a different definition of relationship and therefore relevant to a different social process.

1.2   Web Forum as an OSM

Despite the fact that Social Networking Sites (SNSs)[1] are the most popular online sites [24], there are numerous other ways in which a user participates in OSM activities, e.g., through Web forums, blogs, wikis, bookmarks, diggs, RSS, and so on, where each one has its own distinctive role and effect on the relationship being established. Unlike other OSM[2], *Web forums* or *discussion boards* provide a platform for formal, vivid and dynamic discussions among an unrestricted number of participants. Figure 1 shows a typical ecosystem formed by user interactions in a Web forum. In this folksonomy, discussions are started by its members in the form of a discussion thread with a title and an entry message post. Viewers of this thread annotate their own opinions or replies to the thread and thus the system keeps on evolving as the number of posts grow in the thread. Starting with an equilibrium state of no posts, it goes through a disequilibrium state once a message is posted as a response, in which the community interacts answering the preceding and following messages. It reaches back to equilibrium state after all the commentators finish up presenting their views and no further message is appended to the thread [5]. During the course of discussions, the interactions in the form of replies and responses stir to establish some relationship among the unknown users. In the example of fault in the car mentioned in section 1.1, the thread initiator develops a relationship of trust with those replying to him; however the confidence level of trust depends on the *structure* of interactions (replies) and the *relevance* of the message contents to the thread title and entry post. Nevertheless, the unrestricted ordered growth of intertwined posts in a thread with not much support to identify a *reply-to* relationship makes it extremely complicated to trace its actual interaction

---

[1] SNS interactions are usually casual in nature with short and frequent communications.

[2] Each type of OSM has its own distinctive role and effect on the relationship being established.

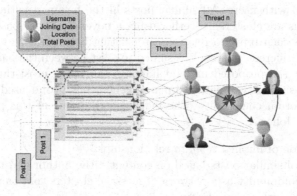

Fig. 1. A Web forum *ecosystem*

structure. After a thread is initiated it starts springing up linearly, but with time as it involves more and more participants, the distinct view of each participant in response to a distinct post very often transforms it to complex multi-threaded structure [27]. It has been found that an interaction structure coordinate with social media analysis in a variety of ways [12], like identifying user roles, their social values and the social community structure [22], establishing *ties* among users [13], and so on. The inherent complexities of thread structures and user interactions as well as lack of functionalities in producing organized information in Web forums have actuated research on tracing interaction among users in a forum [12,25].

1.3   Contributions

A common phenomenon observed in online threaded discussions is that they usually start from a specific topic, but as they grow with more posts, their context goes on deviating from its actual title [15]. Very often a deviated discussion is found to be overlapping with a different thread in the forum. A person replying to the deviated post in one thread is very much likely to reply the similar posts in other threads if he comes to know about this kind of thread overlaps. The state-of-the-art research makes it very clear that the *reply-to* relationships play a prime role in interaction graph generation [12]. But in case of a deviated discussion, a simple reply-to relationship fails to capture the relation between a reply-post in a thread, and the posts in other threads which are similar to the post to which the former is replied. This paper basically addresses the problem of modeling a Web forum ecosystem into a social graph. We propose a novel enriched social graph generation method, which (in addition to identifying reply-to relationships) identifies message-similarity relationship to keep track of all similar posts resulting out of deviated discussions and thus models cross-thread community interactions and interests. The proposed method still considers *reply-to* relations as the primary means of linkage, but rather than establishing links between users, links are established between clusters of similar posts which are in turn associated with users. All similar posts in the forum resulting from deviated discussions are clustered together using a novel similarity-based clustering algorithm, and each reply-to relationship existing between a pair of posts belonging to two different clusters is assumed to exist between the pair of clusters. The novelty of the proposed method lies in establishing cross-thread linkages using the post-similarity[3] relationship, and generating a condensed social graph of the entire forum community. The main contributions of this paper can be summarized as follows.

- identification of implied *reply-to* relationships;
- clustering of similar posts based on content, title, author and time; and
- modeling community as a network of message clusters that can be explored at different levels of specificity.

For this, the method starts with linking posts in each thread individually by identifying reply-to relationships. Thereafter, a clustering algorithm based on similarity of posts is applied across the forum to group all posts into different clusters. Finally, each relation between two individual posts is mapped between the clusters to which these posts belong. Hence, the enriched interaction graph generates a network of clusters that can also be presented in the form of users who authored those posts to generate the social network of users, and at the same time it keeps information for all other users with similar interests. This work is an extension of [2] published in the proceedings of MSM'12.

The rest of the paper is organized as follows. Starting with a review of prior studies on interaction graph generation for Web forums in section 2, we discuss the enriched social graph generation method in section 3. Section 4 presents the experimental setup and evaluation results to establish efficacy of the proposed method including a brief discussion. Finally, section 5 concludes the paper with possible future enhancements.

2   Related Work

Irrespective of the type of social media (SNS, Web log, forum, video-sharing site, etc.) research on characterizing user behaviors and their interaction structures have always been a pioneering area in field of social media analysis [4,25]. Generally, interaction structure among users play an important role in generating relevant social networks from their communications, which in turn can be applied to mine all other related information.

The inherent complexities and lack of support from the online platforms powering forums bring about various challenges in capturing user interaction structures, roles and behaviors. Identifying user roles is a well established problem in Web forum analysis. Himelboim *et al.* [19] analyzed social roles in political forums to distinguish between social leaders and the rest. Chan and Hayes [7]

---

[3] The word "post" appearing throughout the paper refers to its noun form as "message-post", and should not be confused with its adjective form.

established user communication roles in discussion forums by analyzing several different categories of features including structural, reciprocity, persistence, popularity, and initialization.

Gomez *et al.* [14] created a social network from discussion threads in *Slashdot* using user interactions and their main objective remained statistical analysis of the generated network. The Hybrid Interactional Coherence (HIC) algorithm [12] generates an interaction graph of users that is basically composed of reply-to interactions. As reply-to relations are not always explicit in a Web forum, Fu *et al.* adopted three key feature-matches including system feature match (consisting of header information match and quotation match), linguistic feature match (consisting of direct address match and a lexical match algorithm), and residual match. Rather than using the reply-to relationship between posts, Liu *et al.* [25] exploited the similarity measure to generate structure of the social network of a forum. They defined similar people to represent friendship, shared interest, or skill-similarity. For similarity comparison between different posts, they defined a measure that considers post content similarity, thread title similarity, and author similarity. In [22], Kang and Kim generated an information flow network from discussion threads. In their network, a node represents either a user or a message, and an edge represents the reply-to or authorship relationship. Messages posted by same user are connected globally across the forum in different threads using an authorship relationship. Unlike others, Aumayr *et al.* [3] applied a machine learning approach to capture the reply-to relationships using a set of five fundamental features as *reply distance, time difference, quotes, cosine similarity,* and *thread length.* They used SVM (support vector machine) and C4.5 (extension of decision tree based ID3 algorithm) classifiers and comparatively analyzed them by varying the feature combinations. In our earlier work [1], we have applied a similarity-based clustering approach to identify cliques in dark Web forums.

3   Proposed Method

The proposed enriched social graph mining method primarily consists of five major tasks as shown in Figure 2. It starts with forum crawling and parsing to fetch thread data, which is followed by some preprocessing tasks, reply-to relationship identification, similarity-based clustering, and finally, their integration to construct the enriched social graph. Further detail is presented in the following subsections.

3.1   Forum Crawling and Pre-processing

The process starts with data crawling and pre-processing step, in which a URL of the forum homepage is passed to the forum crawler[4] which crawls all the webpages in this domain and eliminates the duplicates heuristically. A platform-specific parser module is employed to extract all the meaningful pieces of information from the crawled webpages, which are passed to the data pre-processing

---

[4] Our crawler is based on `crawler4j` package (http://code.google.com/p/crawler4j/)

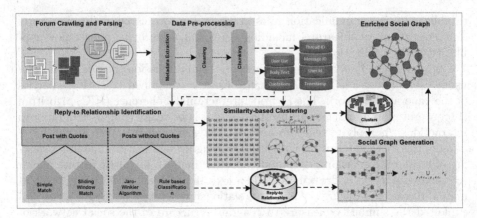

Fig. 2. Proposed social graph mining method

module. The metadata extraction task works in close coordination with the parser module to extract all the metadata. Thus, we get the data organized as a collection of threads having a title and a unique id, each thread consisting of one or more posts that in turn comprises a post id, time-stamp, body, author and quotations, if they exist. Details about each author comprising user id, joining date, location, and total posts are collected separately. The body text is additionally processed by some cleaning and chunking to smoothen its noise and tokenize into individual meaningful pieces of information. The most common form of noise in message posts are the unnecessary repeated use of characters like punctuation marks, symbols, letters, and digits along with letters, e.g., "okkkk". These kinds of noises are dealt by cleaning the body text. Then the body text is divided into different text chunks, called chunking, where boundaries are decided by the punctuation symbols like full-stop, comma, colon, and semicolon. It leads to produce good quality n-grams in subsection 3.3.

3.2 *Reply-to* Relationship Identification

When a thread is initiated by someone, it is assigned a title, and an initial post is attached with it, often called *entry post*. The entry post simply elaborates its title and waits for other's comments on it. Viewers, who find interest on the newly initiated thread, comment on it, 1) either by quoting an existing post to respond specifically, 2) or by a quote-less post. In the first case when somebody quotes a post, the reply-to relationship becomes absolutely clear, but it is just the reverse in the other case. For example, let us suppose a user u_1 initiates a thread and another user u_2 comments on it. If u_2 commented by quoting the post of u_1, then u_2 makes it clear that she is replying to u_1 ($u_2 \Rightarrow u_1$), but if she commented simply without any quote, it remains unclear that to whom did u_2 reply. At the same time, as the post of u_1 is the only post other than her own, it indirectly resolves to establish the relationship as $u_2 \Rightarrow u_1$. After that, suppose u_3 replies in that thread. In case she quotes someone (either u_1 or u_2), the reply-to relationship becomes clear, else it becomes ambiguous and all the relations, $u_3 \Rightarrow u_1$, $u_3 \Rightarrow u_2$ and $u_3 \Rightarrow \{u_1, u_2\}$ have equal probability to exist. In this way, the more a thread grows in length, more ambiguous becomes the reply-to relationship. This section presents an approach to establish the reply-to relationship for each post commented in a thread.

Case 1: Posts with Quotes. Most of the time quotations accompanying a post occur as a simple single quote to another post. Multiple quotes (a post quoting multiple other posts at a time) and nested quotes (a post quoting a quoted post), are also encountered occasionally to focus on specific points in the discussion. All of them are neutralized by breaking down the multiple quotes into multiple single quotes, and processing the nested quotes to drop all the nested inner quotes except the outermost. Other issues regarding quotes are that sometimes a Web forum engine may itself modify the format of quotation and it also provides authority to the author to modify a default quote [28]. An author may sometimes find a lengthy quote message to be cumbersome, and to focus on a specific point may edit the message to delete rest of its body. In this kind of behavior, it becomes difficult to trace the post to which is it responding by the quote. To overcome these issues, if a simple complete match fails to identify a reply-to relation, we follow a sliding window technique [26,12]. In this technique, the text of earlier posts as well as the quote is broken down into substrings (windows) and the quote-post pair with highest number of substring matches are linked.

Case 2: Posts without Any Quote. For comments that are posted without quoting any of the existing posts, because of having no sound clue it becomes very difficult to establish the reply-to ($\Rightarrow$) relationship. Although some prior research works use the notion of similarity of posts to establish a reply-to relationship [12], contradictory to this, we found that simply a similarity of textual contents doesn't provide much evidence for a reply-to linkage. Rather a higher similarity shows an imitation of the same words. For example, let us suppose a thread is initiated with an entry post, p_1, asking for help to learn Java, and a Java expert after noticing p_1, replies in p_2 ($p_2 \Rightarrow p_1$) by explaining some basic concepts of Java. Another Java expert caught attention of this discussion and replied in p_3 to p_1 ($p_3 \Rightarrow p_1$) by explaining some more concepts. Now, in this thread as p_2 and p_3 are explaining on the same topic and p_1 is just a question asking help, p_1 and p_2 have a high probability to be similar even more than p_1, but neither is replying to the other. Thus in this paper, we have differentiated a reply-to relationship from the property of posts being similar.

While commenting in a thread, very often people use author name of an earlier post in text to reply to that specific user, instead of quoting [12]. To capture this information, a search for a match of usernames of earlier posts in the body text may lead to establish an obscured reply-to link. At the same time, as we know that an online conversation is hardly given a serious attention, the writing style remains far from a formal way of writing [18].

Unintentional misspellings and grammatical errors are commonly found in them [17], and many times usernames which do not look like real names are intentionally trimmed to make it like a real name. To overcome this hurdle, we apply an approximate string matching (ASM) algorithm. In [8], Cohen *et al.* performed a comparative study of string distance metrics for name matching and found Jaro-Winkler metric [30] as intended primarily for short strings. In our research study of Web forums, we found almost 90% of usernames to be in single words and the Jaro-Winkler metric suited best for us to match misspelled usernames.

First we define two basic measures used in it. For two strings $s_1 = a_1 \cdots a_k$ and $s_2 = b_1 \cdots b_l$, a character a_i in s_1 is defined to be *common* with s_2 if there is a $b_j = a_i$ in s_2 such that $(i - H) \leq j \leq (i + H)$, where H is calculated using equation 1.

$$H = \frac{min\left(|s_1|, |s_2|\right)}{2} \qquad (1)$$

Now, let us suppose $s_1' = a_1' \cdots a_{k'}'$ be the characters in s_1 which are common with s_2 (in the same order they appear in s_1) and let $s_2' = b_1' \cdots b_{l'}'$ be analogous to s_1'. A *transposition* of s_1' and s_2' is defined to be a position i such that $a_i' \neq b_i'$. The basic Jaro metric [20] measures the similarity, $J(s_1, s_2)$, between s_1 and s_2, using equation 2, where, $t_{(s_1', s_2')}$ is half the number of *transpositions* for s_1' and s_2'.

$$J(s_1, s_2) = \frac{1}{3} \times \left(\frac{|s_1'|}{|s_1|} + \frac{|s_2'|}{|s_2|} + \frac{|s_1'| - t_{(s_1', s_2')}}{|s_1'|} \right) \qquad (2)$$

Based on an observation that most common typographic variations occur towards the end of a string, Winkler [30] enhanced the Jaro similarity function into equation 3, where $P' = max(P, 4)$, P being the number of characters in the longest common prefix in s_1 and s_2.

$$JW(s_1, s_2) = J(s_1, s_2) + \frac{P' \times (1 - J(s_1, s_2))}{10} \qquad (3)$$

The value of JW metric is calculated for each pair consisting of a username from earlier posts and an n-gram in the body text. While computing the values, single word usernames are paired with uni-grams of body text, double word usernames are paired with bi-grams, and so on. A threshold value is used to confirm a match with the misspelled name, and accordingly a reply-to relationship is established with the post authored by the matched username. In case more than one post exists from that user, the relationship is linked with the latest post.

Even after applying username string matching algorithm in the body text, there remains considerable number of reply-to relationships undiscovered, and to identify which we follow a rule based classification. In this matching, we make use of communication patterns as in HIC [12], which are briefed below.

Rule Set. Let x be the residual message of author A, y be the previous message of author A, and Z be the set of all the messages of other authors which are posted between y and x and are replies to messages of author A.

Rule 1: If y does not exist, x replies to the first message in the discussion;

Rule 2: If y exists and Z isn't empty, x replies to all the message posts in Z;

Rule 3: If y exists and Z is empty, x replies to what y replies to.

3.3   Post Similarity Identification

Prior research show that a similarity comparison of Web forum posts is not as trivial as usual content similarity [25]. Liu *et al.* [25] defined this measure as a function of body text appended by thread title and author of the post. In our analysis, we noticed an additional factor to count for the similarity measure. Generally, time plays a substantial role in deciding the topics of discussion and its deviation, with respect to the daily happenings in one's personal life. For example, immediately after the tsunami outbreak in Japan in March 2011, all social media got flooded with this hot discussion all over the world. Hence, we observed that the discussions going in close proximity are likely to be more similar than those with a considerable time gap, and we have incorporated timestamp of a post along with other factors to measure similarity as described here.

To find overall similarity between a pair of posts, we calculate four different similarity measures as *content similarity*, *title similarity*, *author similarity* and *time similarity*. Let $D = \{d_1, d_2, \cdots, d_n\}$ be the set of discussion threads and $P^i = \{p_1^i, p_2^i, \cdots, p_m^i\}$ be the set of ordered posts in thread d_i in a forum F. After being cleaned and chunked in the pre-processing step, each post p_j^i is converted into bag of unigrams, bigrams and trigrams, separately. Those either begining or ending with a stopword are filtered out. We use the vector space model (VSM) to transform each post into vectors of unigrams, $\overrightarrow{Uni}_j^i$, bigrams, $\overrightarrow{Bi}_j^i$, and trigrams, $\overrightarrow{Tri}_j^i$, using their *tf-idf* values. The content similarity $CSim(p_j^i, p_l^k)$ between each pair of posts, p_j^i and p_l^k is calculated using equation 4, where $\alpha_1 \leq \alpha_2 \leq \alpha_3$ are constants such that $\alpha_1 + \alpha_2 + \alpha_3 = 1$.

$$CSim(p_j^i, p_l^k) = \alpha_1 \times \frac{\overrightarrow{Uni}_j^i \cdot \overrightarrow{Uni}_l^k}{\left\|\overrightarrow{Uni}_j^i\right\| \left\|\overrightarrow{Uni}_l^k\right\|} + \alpha_2 \times \frac{\overrightarrow{Bi}_j^i \cdot \overrightarrow{Bi}_l^k}{\left\|\overrightarrow{Bi}_j^i\right\| \left\|\overrightarrow{Bi}_l^k\right\|}$$

$$+\alpha_3 \times \frac{\overrightarrow{Tri}_j^i \cdot \overrightarrow{Tri}_l^k}{\left\|\overrightarrow{Tri}_j^i\right\| \left\|\overrightarrow{Tri}_l^k\right\|} \tag{4}$$

Thread title similarity, $LSim(p_j^i, p_l^k)$ is calculated in the same way as content similarity. The only difference lies in the text content which in this case is the text of thread title, as shown in equation 5.

$$LSim(p_j^i, p_l^k) = CSim(title(p_j^i), title(p_l^k)) \tag{5}$$

Author similarity, $ASim(p_j^i, p_l^k)$, is calculated using equation 6, whereas time similarity $TSim(p_j^i, p_l^k)$ is calculated using equation 7, where, $ts()$ stands for the timestamp of the associated post, and $\beta_1 \in [0, 1]$ is a constant.

$$ASim(p_j^i, p_l^k) = I_{[author(p_j^i)==author(p_l^k)]} \tag{6}$$

$$TSim(p_j^i, p_l^k) = \beta_1^{\left| ts(p_j^i) - ts(p_l^k) \right|} \tag{7}$$

Finally the overall similarity, $Sim(p_j^i, p_l^k) \in [0, 1]$, is defined by aggregating all four measures using equation 8, where α, β, γ and δ are constants such that $\alpha + \beta + \gamma + \delta = 1$.

$$Sim(p_j^i, p_l^k) = \alpha \times CSim(p_j^i, p_l^k) + \beta \times TSim(p_j^i, p_l^k)$$
$$+\gamma \times ASim(p_j^i, p_l^k) + \delta \times LSim(p_j^i, p_l^k) \tag{8}$$

3.4   Thread Post Clustering

Online threaded discussions usually start from a specific topic but as a thread grows with more posts, it's context deviates from its actual title [15]. Very often this deviated discussion is found to be overlapping with another one going on in a different thread. To capture this inter-thread similarity, in this step we follow a cost-effective agglomerative clustering algorithm shown in Algorithm 1 to group all similar posts across the forum. It starts with assigning all the different forum posts in a separate cluster. Let us suppose there are n_0 number of total posts in the forum and at time $t = 0$ it starts with $C^0 = \{c_1^0, c_2^0, \cdots, c_{n_0}^0\}$ as the set of clusters assuming that every post is dissimilar from others. At each iteration, t, in the clustering process, a similarity matrix $\Phi_{n_t \times n_t}^t$ is maintained containing the similarity information between each pair of clusters. For the initial similarity matrix, $\Phi_{n_0 \times n_0}^0$, at $t = 0$ its values are calculated as a similarity measure between each pair of posts as shown in equation 9, where $p_i \in c_i^0$ and $p_j \in c_j^0$.

$$\Phi_{ij}^0 = Sim\,(p_i, p_j) \tag{9}$$

At time, t, each value in the matrix, $\Phi_{n_t \times n_t}^t$, is compared with the similarity threshold value, ϵ. The pair of clusters for whom this value is found to be greater are added to the set of pairs, Λ^t, that need to be merged. After collecting all the cluster pairs that show a sign to get merged, they are ranked by their corresponding matrix values. Starting with the top ranking pair, the two clusters are merged to form a unified cluster and all those pairs in Λ^t containing either of the two sub-clusters are removed from the set. The merging process is continued until Λ^t becomes empty. After the completion of merging, it proceeds to next iteration, $t + 1$, the new set of clusters becomes $C^{(t+1)}$ with number of clusters as $n_{(t+1)} < n_t$, and the new matrix becomes $\Phi_{n_{(t+1)} \times n_{(t+1)}}^{(t+1)}$.

Each cluster, c_i^t, at time, t, keeps information about all its posts grouped into two sub-clusters, $c_k^{(t-1)}$ and $c_l^{(t-1)}$, if c_i^t is a result of merging $c_k^{(t-1)}$ and $c_l^{(t-1)}$, else c_i^t contains a single cluster of posts, $c_k^{(t-1)}$, the same as it was in last iteration. Each value, Φ_{ij}^t, in the new matrix is calculated using equation 10, where $|c_i^t|$ and $|c_j^t|$ denote the number of sub-clusters in c_i^t and c_j^t, respectively.

$$\Phi_{ij}^t = \frac{\displaystyle\sum_{c_k^{(t-1)} \in c_i^t, \, c_l^{(t-1)} \in c_j^t} \Phi_{kl}^{(t-1)}}{|c_i^t| \cdot |c_j^t|} \tag{10}$$

Algorithm 1. Post clustering algorithm

Input: A set of posts $P = \{p_1, p_2, \ldots, p_n\}$
Output: A set of cluster of posts $C = \{c_1, c_2, \ldots, c_m\}$
1   $C^0 = \{c_1^0 \leftarrow p_1, c_2^0 \leftarrow p_2, \ldots, c_{n_0}^0 \leftarrow p_n\}$;
2   $t \leftarrow 0$;
3   **repeat**
4   **if** $t = 0$ **then**
5        $\Phi_{n_t \times n_t}^t \leftarrow createSimilarityMatrix(C^0)$;
6   **else**
7        $\Phi_{n_t \times n_t}^t \leftarrow createMatrix(n_t \times n_t)$;
8   **forall the** i *and* j *in* Φ_{ij}^t **do**
9          $\Phi_{ij}^t = \dfrac{\displaystyle\sum_{c_k^{(t-1)} \in c_i^t, \, c_l^{(t-1)} \in c_j^t} \Phi_{kl}^{(t-1)}}{|c_i^t| \cdot |c_j^t|}$;
10   **if** $\Phi_{ij}^t \geq \epsilon$ **then**
11            $\Lambda^t \leftarrow \Lambda^t \cup \{(c_i^t, c_j^t, \Phi_{ij}^t)\}$;
12      $rank(\Lambda^t)$ on decreasing value of Φ_{ij}^t;
13   **repeat**
14        $\{(c_i^t, c_j^t, \Phi_{ij}^t)\} \leftarrow top(\Lambda^t)$;
15        $merge(c_i^t, c_j^t)$;
16   **forall the** *element* $\{(c_k^t, c_l^t, \Phi_{kl}^t)\} \in \Lambda^t$ **do**
17   **if** $c_i^t = c_k^t$ *or* $c_i^t = c_l^t$ *or* $c_j^t = c_k^t$ *or* $c_j^t = c_l^t$ **then**
18   remove $\{(c_k^t, c_l^t, \Phi_{kl}^t)\}$ from Λ^t;
19   **until** Λ^t *becomes empty*;
20      $t \leftarrow t + 1$;
21      $C^t \leftarrow getClusters()$;
22 **until** $|C^t| = |C^{(t-1)}|$;
23 $C = C^t$;
24 **return** C

After t iterations, when there remains no Φ_{ij}^t value greater than the ϵ, the terminating condition in the algorithm shown in Algorithm 1 becomes true and the final clusters are returned as grouped posts. Some spectacular properties of the proposed clustering algorithm are presented below.

- In this algorithm, we do not need to have pre-decided number of clusters (to be generated finally) as is required in most [31]. Rather this number is determined dynamically by comparing Φ_{ij}^t with $\epsilon \in [0, 1]$.

- A strict hierarchical clustering algorithm suffers from its inability to perform adjustment once a merge or split decision has been taken [16], whereas the proposed algorithm is free from this limitation as before merging we rank the cluster pairs to make sure that the merged cluster would not need to be split up later.
- Due to heavy computations, the cost of clustering usually remains very high [31,16]. The time complexity of the proposed clustering algorithm is tn^2, t being the number of iterations, and in worst case it may go up to n^3. However, during its execution as more and more sub-clusters get merged in successive iterations, the dimension of the similarity matrix decreases and the number of computations reduce heavily to make it an efficient algorithm.

3.5 Enriched Social Graph Generation and the Ecosystem Dynamics

In prior research [14,12,25,3], user interactions in Web forums have been defined to generate a network of users for analyzing their activities in different ways, and the reply-to relationships are identified as most prominent features to trace them. In addition, post similarity has been found as another important feature to define a network of users of similar interests [25]. In HIC [12], similarity among the posts in a thread are used as a heuristic to establish a reply-to relationship. However the proposed web forum ecosystem modeling method differentiates a reply-to relationship from the property of posts being similar. The enriched social graph considers a cluster of similar posts in the forum as a node, and the reply-to relationships between posts from different clusters as directed links to connect the nodes. Let $P = \{p_1, p_2, \cdots, p_m\}$ be the set of total posts in all the threads and $R = \{r_{ij}\}$ be the set of relationships $p_i \Rightarrow p_j$ between posts, p_i and p_j. Let us suppose that the set of clusters generated using the clustering algorithm is $C = \{c_1, c_2, \cdots, c_n\}$. Now, the enriched social graph consists of n cluster nodes with the set of relationships, $R^e = \{r^e_{kl}\}$, where r^e_{kl} is defined in equation 11.

$$r^e_{kl} = \bigcup_{p_i \in c_k, p_j \in c_l} r_{ij} \qquad (11)$$

Each post associates with it the thread title, author name or user id and timestamp, and this enriched social graph can be presented in various forms for analyzing the interactions and associations in the ecosystem. When the graph is presented for authors of the posts, it generates a network of authors resembling the *ecosystem actors* connected to others by the reply-to relationship, and at the same time shows all existing actors with similar interest being in the same cluster. An actor with diverse interests may exist in multiple clusters, which shows the diverse nature of her interests. When the clusters are represented by keywords of posts resembling the *ecosystem environment components*, a relation between the topics on discussion issues (or ecosystem environment components) can be identified. When clusters are represented by the timestamp along with keywords of posts, a contemporary network of the hot discussion topics can be generated which characterizes the *ecosystem evolution*. Hence, the generated enriched social graph is actually a multi-purpose graph that captures the distinct features and interactions of a Web forum ecosystem and can be very fruitful for an exhaustive analysis.

4   Experimental Results

The experiments are conducted on a real time test bed described in subsection 4.1. Firstly, the crawler module based on `crawler4j` and parser module developed for the `vBulletin` platform crawls and parses the forum webpages to extract all the meaningful pieces of information from them, which are then passed on to subsequent modules. The proposed social graph generation method is evaluated in different perspectives in subsection 4.2.

4.1   Test Bed

For a realtime test bed, we considered *"eActivism and Stormfront Webmasters"* forum under the *Activism* category in the popular Stormfront[5] social Web forum. As of 04th *Feb.* 2012, this sub-forum consisted of $1,698$ threads getting a total of $11,210$ replies and $2,271,494$ views. Set up in 1995 by Don Black, it is considered by many as a neo-Nazi Web forum and was identified as the first major hate-site on the Web, which drove us to study user interactions in it.

4.2   Experiments

The first major task is to identify the reply-to relationships in-between posts in a thread, that lay a foundation of the proposed graph generation method. This task is evaluated by using the metrics, precision (π), recall (ρ) and F-score (F_1). Precision is defined as the ratio of no. of correctly identified relations to total no. of identified relations, recall is defined as the ratio of no. of correctly identified relations to the no. of relations that actually exist, and F-score is harmonic mean of the duo.

As the metrics needed a gold standard to compare with, some agelong threads relevant to the forum category are shortlisted to manually set its gold standard. Only 29 threads are found having more than 40 comments on them. Discarding the irrelevant ones, we stuck to 10 threads. Two independent users are assigned to manually identify all actual reply-to relationships based on their context, and finally conflicts were resolved on a mutual consent. Another set of relationships identified by the proposed method are also collected. Values of the evaluation metrics are calculated using these two sets, shown in Table 1 along with their statistical summary. In these 10 threads, we see that the F-score value reached as high as 0.900 for thread no. 10 and as low as 0.711 for thread no. 6. The average values for precision, recall and F-score are found as 0.799, 0.815 and

---

[5] http://www.stormfront.org

Table 1. Result summary of *reply-to* relationship identification process

Thread No.	Posts	Participants	π	ρ	F_1
1	68	22	0.784	0.816	0.800
2	105	13	0.727	0.715	0.721
3	122	37	0.831	0.847	0.839
4	82	54	0.802	0.790	0.796
5	185	48	0.878	0.845	0.861
6	58	41	0.691	0.733	0.711
7	55	11	0.856	0.862	0.859
8	169	52	0.773	0.816	0.794
9	44	11	0.758	0.809	0.783
10	46	37	0.887	0.914	0.900
Avg.	**93.4**	**32.6**	**0.799**	**0.815**	**0.806**

0.806 respectively. Out of the two cases mentioned in subsection 3.2 to capture the relationship, the first case was found to be accurate with 0.987 as its average F-score. However in the second case, due to its high ambiguity, accuracy of the system went down to an average F-score of 0.641.

As another part of this experiment, the established relationships between posts are transformed to establish the same relations between users. As there exist some common users in different threads, the relationship established for a user in one thread is continued over and integrated with the relationships in other threads, which connected users in different threads to form a network. The generated network consisted of 3 distinct components, each of whom represent a closed-group of inter-related users, shown in Figure 3. There are a total of 310 nodes (overall participants) and 545 directed edges (reply-to relationships) generated from a total of 934 posts distributed in 10 discussion threads. Figure 4 presents a zoomed view of the two small components.

In the next experiment, we considered to test the clustering algorithm based on the designed post similarity measure. It is evaluated in terms of the metrics $F_{\alpha=0.5}$ (or F_P[6] at $\alpha = 0.5$) and $F_{B\text{-}cubed}$ (or F_B[7]). As like the previous experiment, posts are manually grouped by two independent users and conflicts are resolved on mutual consent, which produced a set of 207 clusters as a gold standard from the same set of 934 posts. The proposed method is started with extracting all the unigrams, bigrams, and trigrams. For the constants, α_1, α_2 and α_3, their values are determined by solving the equations, $\alpha_3 = 3 \times \alpha_1$, and $\alpha_2 = 2 \times \alpha_1$, where $\alpha_1 + \alpha_2 + \alpha_3 = 1$ is the condition applied. Thus their

---

[6] If C is the set of clusters generated by the automated system and L is the gold standard set, then $purity = \sum_i \frac{|C_i|}{n} maxPrecision(C_i, L_j)$ and $inversepurity = \sum_i \frac{|L_i|}{n} maxPrecision(L_i, C_j)$. F_P is calculated as their harmonic mean.

[7] For each element (or post), i, precision and recall values are computed individually as $precision_i = \frac{C_i \cap L_i}{C_i}$ and $recall_i = \frac{C_i \cap L_i}{L_i}$. The average b-cubed precision and recall are computed as the mean of individual values. F_B is the calculated as their harmonic mean.

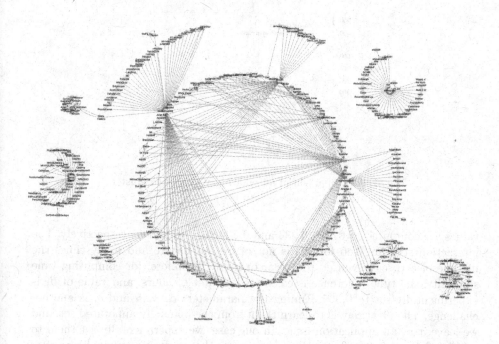

Fig. 3. Network generated by system-identified *reply-to* relations between users

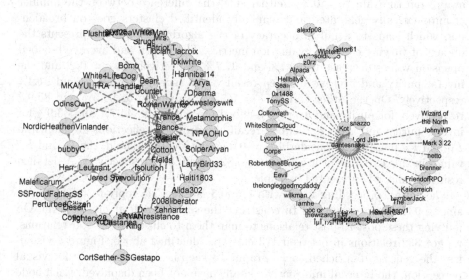

Fig. 4. Zoomed view of the two small components in Figure 3

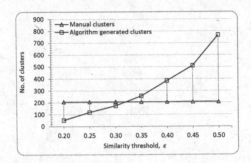

Fig. 5. A comparison of automatically generated clusters

values are computed as 0.167, 0.333 and 0.5 for α_1, α_2 and α_3 respectively. The interaction style in Web forums is not of instant nature and many times the lifetime of a thread even go to a year or more. Therefore, for computing time similarity, the time difference is calculated in unit of hours, and value of β_1 is experimentally set to 0.995. Tuning the parameters α, β, γ and δ, is another challenge. The ideal way is to learn them from the manually annotated set, and we leave it as an application issue. In our case, we experimentally set them to 0.7, 0.1, 0.1 and 0.1, respectively, and generate the similarity matrix. Thereafter the clustering algorithm is executed by varying similarity threshold, ϵ, from 0.2 to 0.5 in intervals of 0.05. The line chart shown in Figure 5 presents the trend of increasing number of generated clusters as the value of ϵ increases. As we move away from the value $\epsilon = 0.3$ in either side, the difference between the number of automatically generated and manually identified clusters goes on broadening, which leads to a fall in accuracy of the algorithm. Figure 6 presents the impact of varying ϵ on the evaluation metrics. The purity and average b-cubed precision values decrease to 50.7% and 37.7% respectively at $\epsilon = 0.2$, and the inverse purity and average b-cubed recall values increase to 86.4% and 91.4% respectively. On the other end, at $\epsilon = 0.5$ the values go to 90.5% and 85.9% respectively for purity and average b-cubed precision, and 44.7% and 33.9% for inverse purity and average b-cubed recall. Accordingly reflections are shown in F_P and F_B measures. Considering $\epsilon = 0.3$ as the ideal threshold, the F_p and F_B values in this experiment are found as 0.825 and 0.804, respectively. A detailed result summary is presented in Table 2.

Thereafter, proceeding forth with $\epsilon = 0.3$, all the 545 reply-to relationships among 934 posts are unified to construct the social graph at cluster-level. On unifying these post-to-post relations to map them to cluster-to-cluster relations, we got 332 relations in-between 173 clusters, identified above. Figure 7 visualizes the generated enriched social graph. To keep it simple and easy for visual perception, the internal informative details have not been displayed. Each node in it is a cluster of posts that are highly similar to each other and each link is a directed reply-to relationship between the two clusters. We see that it consists of two disconnected components, which shows that the posts in one component

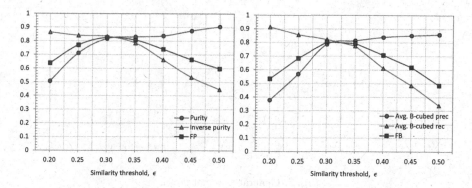

Fig. 6. Impact of ϵ on *Purity* measures and *B-Cubed* measures

Table 2. Result summary of similarity-based clustering algorithm

Metric	Value	Metric	Value
Purity	0.817	Avg. B-cubed Prec	0.787
Inverse Purity	0.834	Avg. B-cubed Rec	0.822
F_P	0.825	F_B	0.804

is neither similar to those in the other, nor the posts in one component replied to any post of the other component. Thus, the small component is either a single thread or a group of very few threads whose topic of discussion is totally different from that going in rest of the threads. As can be seen in the larger component that few nodes are thickly connected to others while most are very thinly connected. The set of posts in thickly connected nodes are getting more attention from other members for the inbound links and their outbound links show that their authors are active members in the forum. Irrespective of being inbound or outbound, the thickness of linkages indicates that the topics of posts in the cluster are among the hot issues of that time. The constructed enriched social graph can be used to present the network in multiple ways as described in section 3.5 to present and analyze the dynamics of the web forum ecosystem. The thickness of linkages in the social graph characterizes the kind of ecosystem. A social graph with thick inter-cluster collaborations and linkages create a dense network that characterizes a *strong* ecosystem, whereas a social graph with thin inter-cluster linkages create a sparse network that characterizes a *weak* ecosystem.

Although no experiment is performed on Q/A Web forums such as cross-validated[8] or stack-overflow[9], the methodology is highly applicable to them and such other specific forums. It would start with mining the enriched social graph from the complete Q/A forum and keep on updating the graph at

---

[8] http://crossvalidated.com/
[9] http://stackoverflow.com/

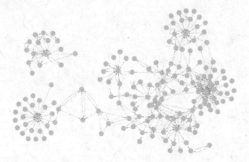

Fig. 7. Generated social graph

regular intervals with the addition of threads and posts in the forum. It could then be used for automatic spontaneous query-answering using the knowledge stored in the graph.

5   Conclusion and Future Work

In this paper, we have considered a Web forum as an ecosystem and presented a methodology to model the discussions into an enriched social graph using user interactions and their overlapping interests, with a deliberate consideration of deviated discussions. The user interactions link posts through *reply-to* relationships, whereas the overlapping interests lead to merge similar posts into clusters, and thus collapse the generated network. Authors of posts in the same cluster share common interests and are linked with people of distinctive interests through the *reply-to* tie. The enriched social graph can serve for analysis of Web forum discourse in multiple ways that can be explored to bring various undiscovered facts. It is also applicable to Q/A Web forums and such other specific forums for automatic query-answering spontaneously.

This work mainly focuses on the approach to generate social graph to model user interactions and their overlapping interests, rather than analyzing forums using it. Therefore, the most important future direction is to devise approaches to analyze Web forums using the generated social graph. There are few issues regarding enhancements of the proposed approach. The linguistic analysis of posts can further be enriched to improve the F-score value of *reply-to* relation identification process for the posts which do not include quotes. Moreover, usually people in real life are tied together by several other kinds of social relationships, which somewhat also exist on the Web. Identification and incorporation of all such relationships in the social graph, along with user activities and behaviors, are good candidates for future work.

References

1. Anwar, T., Abulaish, M.: Identifying cliques in dark web forums- an agglomerative clustering approach. In: Proc. of the 10th IEEE Int'l Conf. on ISI, pp. 171–173 (2012)
2. Anwar, T., Abulaish, M.: Mining an enriched social graph to model cross-thread community interactions and interests. In: Proc. of the 3th Int'l Workshop on MSM, Co-located with 23rd ACM Int'l Conf. on HT, pp. 35–38 (2012)
3. Aumayr, E., Chan, J., Hayes, C.: Reconstruction of threaded conversations in on-line discussion forums. In: Proc. of the AAAI ICWSM, pp. 26–33 (2011)
4. Benevenuto, F., Rodrigues, T., Cha, M., Almeida, V.: Characterizing user behavior in online social networks. In: Proc. of the 9th ACM SIGCOMM Internet Measurement Conf., pp. 49–62 (2009)
5. Bentivoglio, C.A.: Recognizing community interaction states in discussion forum evolution. In: AAAI Fall Symposium Series, pp. 20–25 (2009)
6. Brewington, B.E., Cybenko, G.: How dynamic is the web? Comput. Netw. 33(1-6), 257–276 (2000)
7. Chan, J., Hayes, C., Daly, E.: Decomposing Discussion Forums using User Roles. In: Proc. of the WebSci 2010: Extending the Frontiers of Society On-Line (2010)
8. Cohen, W., Ravikumar, P., Fienberg, S.: A comparison of string distance metrics for name-matching tasks. In: Proc. of the Int'l Workshop on IIWeb, held with IJCAI, pp. 73–78 (2003)
9. Correa, T., Hinsley, A.W., de Zúñiga, H.G.: Who interacts on the web?: The intersection of users' personality and social media use. Comput. Hum. Behav. 26(2), 247–253 (2010)
10. De Choudhury, M., Mason, W.A., Hofman, J.M., Watts, D.J.: Inferring relevant social networks from interpersonal communication. In: Proc. of the 19th Int'l Conf. on WWW, pp. 301–310 (2010)
11. El Abaddi, A., Backstrom, L., Chakrabarti, S., Jaimes, A., Leskovec, J., Tomkins, A.: Social media: source of information or bunch of noise. In: Proc. of the 20th Int'l Conf. Companion on WWW, pp. 327–328 (2011)
12. Fu, T., Abbasi, A., Chen, H.: A hybrid approach to web forum interactional coherence analysis. J. Am. Soc. Inf. Sci. Technol. 59(8), 1195–1209 (2008)
13. Gilbert, E., Karahalios, K.: Predicting tie strength with social media. In: Proc. of the 27th Int'l Conf. on Human Fact. in Comp. Sys., pp. 211–220 (2009)
14. Gómez, V., Kaltenbrunner, A., López, V.: Statistical analysis of the social network and discussion threads in slashdot. In: Proc. of the Int'l Conf. on WWW, pp. 645–654 (2008)
15. Guan, Y.-H., Tsai, C.-C., Hwang, F.-K.: Content analysis of online discussion on a senior-high-school discussion forum of a virtual physics laboratory. Instructional Science 34(4), 279–311 (2006)
16. Han, J., Kamber, M., Pei, J.: Data Mining: Concepts and Techniques, 2nd edn., pp. 408–418. Morgan Kaufmann (2006)
17. Hargittai, E.: Hurdles to Information Seeking: Spelling and Typographical Mistakes During Users' Online Behavior. J. of the Assoc. for Information Systems 7(1), 52–67 (2006)
18. Herring, S.C.: Computer-mediated communication on the internet. Ann. Rev. of Inf. Sc. and Tech. 36(1), 109–168 (2002)
19. Himelboim, I., Gleave, E., Smith, M.: Discussion catalysts in online political discussions: Content importers and conversation starters. J. of Computer-Mediated Comm. 14(4), 771–789 (2009)

20. Jaro, M.A.: Advances in Record-Linkage Methodology as Applied to Matching the 1985 Census of Tampa, Florida. J. of the Am. Statistical Assoc. 84(406), 414–420 (1989)

21. Jones, S., Fox, S.: Generations online in 2009. Technical report, PewResearch Center (2009),
    http://www.pewinternet.org/Reports/
    2009/Generations-Online-in-2009.aspx

22. Kang, J.-H., Kim, J.: Analyzing answers in threaded discussions using a role-based information network. In: Proc. of the 3rd IEEE Int'l Conf. on Soc. Comp. (2011)

23. Kaplan, A.M., Haenlein, M.: Users of the world, unite! the challenges and opportunities of social media. Business Horizons 53(1), 59–68 (2010)

24. Lenhart, A.: Adults and social network websites. Technical report, PewResearch Center (2009),
    http://www.pewinternet.org/Reports/2009/
    Adults-and-Social-Network-Websites.aspx

25. Liu, D., Percival, D., Fienberg, S.E.: User interest and interaction structure in online forums. In: Proc. of the 4th Int'l AAAI Conf. on Weblogs and Soc. Med., pp. 283–286 (2010)

26. Nahnsen, T., Uzuner, O., Katz, B.: Lexical chains and sliding locality windows in content-based text similarity detection. Technical report, MIT (CSAIL), MIT-CSAIL-TR-2005-034, AIM-2005-017 (2005),
    http://dspace.mit.edu/handle/1721.1/30546

27. Rosé, C.P., Di Eugenio, B., Levin, L.S., Carol: Discourse processing of dialogues with multiple threads. In: Proc. of the 33rd Ann. Meet. on Assoc. for Comp. Ling., pp. 31–38 (1995)

28. Severinson Eklundh, K.: To quote or not to quote: Setting the context for computer-mediated dialogues. Language@Internet 7(5) (2010)

29. van Dijck, J.: Users like you? theorizing agency in user-generated content. Media Culture Society 31(1), 41–58 (2009)

30. Winkler, W.E.: String comparator metrics and enhanced decision rules in the fellegi-sunter model of record linkage. In: Proc. of the Section on Survey Research, pp. 354–359 (1990)

31. Xu, R., Wunsch, D.: Survey of clustering algorithms. IEEE Trans. on Neural Networks 16(3), 645–678 (2005)

Author Index